Nature-Inspired Algorithms

This comprehensive reference text discusses nature-inspired algorithms and their applications. It presents the methodology to write new algorithms with the help of MATLAB® programs and instructions for a better understanding of concepts. It covers well-known algorithms including evolutionary algorithms, genetic algorithms, particle swarm optimization, differential evolution, and recent approaches, including grey wolf optimization. A separate chapter discusses test case generation using techniques such as particle swarm optimization, genetic algorithms, and differential evolution algorithms.

The book

- discusses in detail various nature-inspired algorithms and their applications;
- provides MATLAB programs for the corresponding algorithm;
- presents a methodology to write new algorithms;
- examines well-known algorithms like the genetic algorithm, particle swarm optimization and differential evolution, and recent approaches like grey wolf optimization;
- provides conceptual linking of algorithms with theoretical concepts.

The text will be useful for graduate students in the field of electrical engineering, electronics engineering, computer science and engineering.

Discussing nature-inspired algorithms and their applications in a single volume, this text will be useful as a reference text for graduate students in the field of electrical engineering, electronics engineering, computer science and engineering. It discusses important algorithms including deterministic algorithms, randomized algorithms, evolutionary algorithms, particle swarm optimization, Big Bang–Big Crunch algorithm, genetic algorithm, and grey wolf optimization algorithm.

Nature-Inspired Algorithms

For Engineers and Scientists

Krishn Kumar Mishra

CRC Press
Taylor & Francis Group
Boca Raton London

CRC Press is an imprint of the
Taylor & Francis Group, an **informa** business

First edition published 2023
by CRC Press
6000 Broken Sound Parkway NW, Suite 300, Boca Raton, FL 33487–2742

and by CRC Press
4 Park Square, Milton Park, Abingdon, Oxon, OX14 4RN

CRC Press is an imprint of Taylor & Francis Group, LLC

Library of Congress Cataloging-in-Publication Data
Names: Mishra, K. K. (Professor of Computer Science), author. Title: Nature-inspired algorithms : for engineers and scientists / K.K. Mishra.
Description: First edition. | Boca Raton : CRC Press, [2023] | Includes bibliographical references and index.
Identifiers: LCCN 2022009541 (print) | LCCN 2022009542 (ebook) | ISBN 9780367750497 (hbk) | ISBN 9781032322643 (pbk) | ISBN 9781003313649 (ebk)
Subjects: LCSH: Nature-inspired algorithms. | Engineering—Data processing. | Science—Data processing.
Classification: LCC QA76.9.N37 M57 2023 (print) | LCC QA76.9.N37 (ebook) | DDC 571.0284—dc23/eng/20220527
LC record available at https://lccn.loc.gov/2022009541
LC ebook record available at https://lccn.loc.gov/2022009542

ISBN: 978-0-367-75049-7 (hbk)
ISBN: 978-1-032-32264-3 (pbk)
ISBN: 978-1-003-31364-9 (ebk)

DOI: 10.1201/9781003313649

Typeset in Times
by Apex CoVantage, LLC

Contents

Preface

Nature-Inspired Algorithms: For Engineers and Scientists was written to explain the basic concepts of designing nature-inspired algorithms. It also presents how these algorithms can be used to solve science and engineering-related problems.

Nature-inspired algorithms are randomized algorithms that are used for solving complex optimization problems. In addition to solving optimization problems, these algorithms also help solve prediction-related problems and machine learning problems. These algorithms are designed by mapping some natural phenomena in the form of a computer program. Algorithms that are taken in this book are related to optimization problems. After reading this book, a reader will be able to design optimization algorithms and use these algorithms to solve real-life problems. In general, two types of nature-inspired algorithms are trendy. The first category is known as the evolutionary algorithm. Another category is devoted to swarm intelligence-based algorithms.

Evolutionary algorithms have been created by mapping the theory of natural selection and evolution. One such effort was made by prof John Holland, who first introduced a genetic algorithm program to solve optimization problems. A genetic algorithm is a very popular algorithm, and it can be used for solving both discrete and continuous optimization problems. In addition to the genetic algorithm, several other optimization algorithms have also been designed by mapping the theory of natural selection and evolution. These algorithms are called evolutionary algorithms. Examples of some popular evolutionary algorithms are evolutionary strategy, evolutionary programming, genetic programming, and improved environmental adaptation method.

Some optimization algorithms are created by implementing swarm intelligence techniques in the form of a computer program. The swarm intelligence theory has been used to frame many optimization algorithms like Particle Swarm Optimization (PSO), Grey Wolf Optimization (GWO), whale optimization, Bat Algorithm, Ant Colony Optimization, Bee Colony, and many more. It was not possible to cover each algorithm in detail. I have discussed some popular algorithms like PSO and GWO while others are explained in short.

Nature-Inspired Algorithms is an advanced course, and it is offered as an elective subject in B.Tech, M.Tech, and PhD courses at almost all universities. Sometimes in place of the nature-inspired algorithm, only a subset of essential algorithms are taught, and the same course is offered by other names, such as Genetic Algorithms, Evolutionary Algorithms, or Advanced Evolutionary Algorithms. This course is offered as both a graduate or an undergraduate-level course.

Because it is impossible to collect all nature-inspired algorithms, both classical and advanced, in one book, only a subset of algorithms is compiled to create some books on nature-inspired algorithms. In addition to nature-inspired algorithms, some good books have been written on evolutionary algorithms and swarm intelligence algorithms. However, books that cover both algorithms and their applications are uncommon. This book covers both the algorithm and the application of nature-inspired algorithms. This book begins with the fundamentals and progresses through eleven chapters, with the final three devoted to the applications of nature-inspired algorithms.

I drew on our years of teaching experience to address students' persistent difficulties when learning this subject. The book is written with fair assumptions from the student in mind, so it contains more arguments to help elucidate the concept as clearly as possible. A person beginning research in this area often finds it difficult to determine the starting point of his groundwork. In Chapters 8, 10, and 11, I have provided a simplified version of advanced research topics to encourage students to research this subject.

This book has eleven chapters. Chapter 1 covers the necessary preliminaries and provides insight into the subject's fundamental tenets. Chapters 2 and 3 explain various versions of genetic algorithms. The theory of the differential evolution algorithm is covered in Chapter 4. Chapter 5 introduces the particle swarm optimization algorithm. Chapter 6 focuses on the grey wolf optimization algorithm. Environmental adaptation methods and their variants are discussed in Chapter 7. In Chapter 8, you'll learn about some more important nature-inspired algorithms. Applications of nature Inspired algorithms are demonstrated in Chapters 9, 10, and 11.

Almost every chapter in this book contains a rich and extensive pedagogy. Practicing is the best way to become acquainted with the topics embodied in a typical course on nature-inspired algorithms. As a result, at the end of each chapter, I've included multiple-choice and theoretical questions.

I hope the students make good use of this book to learn the subject. I also believe that the course instructors will benefit from this book as they discuss the matter in class. I have made every effort to make this book as error-free as possible, and I welcome feedback and discussion on the topics covered in the text.

Acknowledgments

I am very thankful to the authorities of Motilal Nehru National Institute of Technology (MNNIT) Allahabad, Prayagraj, and the members of the Department of Computer Science, University of Missouri, Saint Louis (USML), USA, for enabling me to write this book. I also want to thank my research scholars and M.tech Students, especially Satya Deo, Ravi Prakash, Bhavna Sharma, and Ashish Tripathi, who supported me in writing and reviewing the content of the book. I am also thankful to my colleges and my friends (Shailesh Tiwari, Deepak Kumar Singh, Sandeep Harit, and Akash Punhani) for their encouragement and constant support. I am indebted to Prof. Sanjiv Bhatia, UMSL, who was very helpful in preparing the manuscript. I thank my publisher, CRC Press, for its role in bringing out the book in its present form. In particular, I thank Gauravjeet Singh Reen and Lakshay Gaba for their editorial inputs. I am also thankful to Gaurav and Vishal for helping me in finalizing the formatting of the book.

Last but not the least, I gratefully place on record the love and affection of my family members (Sanyogita, Ambesh, and Pragya).

About the Author

Krishn Kumar Mishra is presently working as an assistant professor, Department of Computer Science and Engineering, MNNIT Allahabad, India. His research areas include genetic algorithms, the analysis of algorithms, automata theory, and microprocessor and multi-objective optimization. He has taught courses including computer architecture, data structures, advanced computer architecture, programming in C++, and microprocessor and automata theory at the undergraduate and graduate levels. He is a regular reviewer of *Journal of Supercomputing* (Springer), *Applied Intelligence, Applied Soft Computing, IEEE Transaction on Cybernetics, IEEE System Journal, Neural Computing and Application*, and *Institution of Electronics and Telecommunication Engineers* journals.

1 Introduction

1.1 INTRODUCTION

We aim to design systems that can produce meaningful output for society on any subject. These systems help solve problems related to science, engineering, medicine, or any other field. Problem-solving is implemented by collecting data from a variety of sources and converting it into valuable information. For the maximum utilization of a system, we need to identify a combination of inputs that can produce the maximum benefit to society. However, as the capacity of these systems is limited, it is quite impossible to determine the value of optimal inputs by checking all possible combination of inputs manually (by the system). Since these systems are created with the help of physical components, which have limited processing capabilities and rely on existing hardware technologies, their processing capability is limited.

If they will be used for identifying optimal input, they might crash or will take huge time for producing the optimal solution. Due to these problems, hardware methods are not useful and should not be used. These methods can be used only when we have to test very few inputs. In complex problems, we have to use some other methods to identify the value of optimal inputs.

Also we want to specify that a system, which is producing maximum benefit to society, can achieve it by minimizing or maximizing the output of the system. The goal of designing each physical system may be different. For example, some systems are designed to reduce the system's output, while others are designed to increase it. Regardless of their goals, these systems are always prepared to serve society in some way.

In some cases, the minimum output value may be the most beneficial to the individual or community. In other cases, however, the most important output value may not be the most useful. To maximize such physical systems' utilization, we have to design an optimization problem that helps identify the system's optimal value by producing the least or most output.

Some example systems and optimization problems:

An optimization problem is a well-known problem in which the goal is to identify the optimum solution that minimizes or maximizes the output of a system. An optimization problem is challenging to solve. It requires a substantial amount of effort. In this chapter, we look at various physical systems and try to create optimization problems related to them. Many algorithms can be designed to solve optimization problems. However, these approaches are dependent on the complexity of the optimization problem. Some of them can be solved manually, while others require specialized algorithms. We discuss these approaches one by one and explain when they can be applied.

Let us start with the simplest method and see if we can solve the optimization problem manually. To manually check a system's optimum value, we must manually

DOI: 10.1201/9781003313649-1

inspect each input and its result. We must identify which output is the highest or lowest after collecting all the output values. Finally, determine which input produces the highest or lowest value to find the optimum value. Although this strategy appears to be simple, it is not viable to implement. It is impossible to check the output value on all conceivable inputs on a physical system due to lack of time and resources. However, this method can be employed when the number of inputs is limited and can be checked physically.

Complex optimization problems can be solved using other optimization methods, including mathematical or nature-inspired approaches. However, these approaches necessitate the representation of optimization problems as an objective function. This objective function is expressed with the help of a mathematical function that explains the relationship between the physical system's input and output. It also includes the input domain's details and the problem's aim (minimization or maximization).

We will take few examples of physical systems and see how we can create an optimization problem and solve it.

We take a simple example of a physical system and see how it can convert input and output. What will be the input and output, and how will these values be connected? What will be the objective function?

Example system (car): A car is an example of a physical system. It receives fuel (input) from the fuel tank and produces energy (distance as output) as an output that is used to move a car from one place to another. It can be checked that capacity of the petrol tank is limited and bounded. Similarly, the energy produced by the engine is dependent on the amount of fuel. The energy produced by the engine is calculated using the distance traveled by car. Let us want to use this system to cover the maximum distance in one go (objective). In this example, we want to identify the amount of petrol that can fulfil our objective to maximize the distance. Graphically the system can be represented by Figure 1.1.

Input to the system: In this example, the input identification is straightforward, and it will be the amount of fuel in the fuel tank. The input values lie between *mf* and *max*, where *mf* is the amount of minimum fuel in the tank and *max* is maximum fuel, represented by the following constraints. The fuel x must satisfy the constrained listed in Equation 1.1, where *mf* and *max* will be the minimum and maximum capacity of the tank.

$$mf \leq x \leq max \tag{1.1}$$

For example, we can represent the input domain as $1 \leq x \leq 30$, where 1 liter is the minimum fuel(*mf*) and 30 liter is the maximum fuel (*max*).

FIGURE 1.1 Example of physical system.

The output of the system: The distance traveled by car will be the output.

Objective: Maximization

Objective function: To apply numerical optimization techniques, we have to define the objective function for this problem. This objective function contains the detail of the mathematical function that describes the relationship between the input and the output of the system and the detail of the input domain and the goal of the objective function. For the given example of a car shown in Figure 1.1, the mathematical function is used to define the relationship between the amount of fuel and the distance traveled by car. Since the distance travelled by car is proportional to the amount of fuel. This relationship between input and output can be defined by the function shown in Equation 1.2, where x and the distance travelled, $f(x)$ represents the distance covered by the car, and c is a constant.

$$f(x) = c \times x \tag{1.2}$$

Representation of optimization problem: Overall, the optimization problem can be represented by Equation 1.3:

$$Max\, f(x) = c \times x, \quad \text{where}\, mf \le x \le max \tag{1.3}$$

Solution of the problem: From the function, it is clear that there is a linear relationship between input and output, and the output value increases as we change the input. By checking the trend (nature) of the mathematical function, we can define that the optimal value of the function will be equal to the max value of fuel in the system.

Let us take a closer look at this procedure. Assume the tank's capacity is between 1 and 30, and the c used in the function is 10. To determine which input value will yield the highest output, we must examine the result for each value of input. However, because the input domain is continuous, x can have an infinite range of deals from 1 to 30. Due to time and resource constraints, it is nearly impossible to check and compare all output values. As a result, we must find a different solution to this problem. We will try to answer close to the optimal global solution by minimizing the total number of evaluations in another method. We will calculate the system's output on some specific inputs to reduce the number of assessments. Then, by examining the function's nature, we can predict the input value to maximize the system's output. Let us decide to measure the distance traveled by car at $x = 2$ and 3. The system's output at 2 units of x is 20. Similarly, when x is 3, the output of the function is 30. By examining these output values, we can see that increasing the amount of fuel increases the distance traveled. Assuming that the function will continue to follow the same trend until the last input, the optimal intake is 30.

Defining solution for complex optimization problems: Identifying the optimal value of the function is simple in some optimization problems, such as maximizing the distance traveled by car discussed in the previous paragraph. However, these methods are not suitable for all optimization problems. They necessitate the use of standard mathematical techniques designed to solve optimization problems.

The mathematical function defining the relation between input and output in complex physical systems may be very complicated. In such problems, these simple mathematical techniques will not help produce an optimal solution. For solving these problems, the application of a generalized approach is required. These new generalized techniques with more computational power can be created by defining new algorithms in a computer program. The optimal value produced by these methods provides a good approximation of the input vector, which can produce optimal output for the physical system.

Let us take one example of a complex problem.

Sugar monitoring system: For the betterment of the country, its citizens' health is of the utmost priority. We are always interested in designing efficient systems that can help patients and minimize the loss of life. One widespread disease in Indian society is diabetes. A patient with diabetes requires proper monitoring of glucose levels in the blood because their glucose level may go up or down very rapidly. These irregular changes in blood glucose level may affect the organs of the diabetic patient and can lead to the patient's early death. To save the life of a diabetic patient, we can design a system that monitors the blood sugar levels and controls sugar levels in the desired range. Such kinds of systems will be beneficial for society. The sugar level in a diabetic patient may be high or low. So different kinds of treatment may be required to increase or decrease the sugar level of the patient. To increase the chances of survival, we can monitor the patient's sugar level, and if it goes down, we can provide medicine to increase it up to the desired maximum or vice versa.

For this complex physical system, many optimization problems can be designed. What we want to optimize in the system depends on our goal. If the sugar level of a diabetic patient is low, then we have to give them medicine to increase their sugar level to the acceptable level. Similarly, if their sugar level is high, then we have to give him/her medicine to decrease their sugar level to the acceptable level.

Input to the system: In this example, the selection of input parameters is not easy. We have to check several parameters and see how they are affecting the sugar level of the patient. For example, the amount of drug is one of the parameters. However, it is dependent on the physical state of the patient, like body weight, height, and body mass index.

The output of the system: Sugar level

Objective: Maximization/Minimization

Objective function: To apply numerical optimization techniques, we have to define the objective function for this problem. This objective function contains the detail of the mathematical function that describes the relationship between the input and the output of the system and the detail of the input domain and goal of the objective function. For the given example, a mathematical function is used to define the relationship between the input parameters and the sugar level. This function can be designed by collecting patient's information and then using some mathematical model for it.

This mathematical function and the details of the input domain as well as the goal of the optimization problem will create an objective function.

Solution: A simple method will not work in such a complex problem, so we have to apply specialized algorithms.

We have seen some examples of physical systems and created optimization problems for them. Several other famous optimization problems are associated with other systems and have been solved by many mathematicians. For example, many mathematical solutions exist for well-known optimization problems, such as the traveling salesperson problem and the transportation problem. These algorithms, however, cannot be applied to other optimization problems. We are always looking for generic solutions that can be used for a wide range of optimization problems. To understand the crux behind optimization problems and their solutions, we have to study them in detail.

This chapter explains these details by formally defining the definition of optimization problems and the details of specialized optimization techniques used to solve these problems. Each of these algorithms may or may not apply to a set of optimization problems some are problem-specific and cannot be applied to other optimization problems. Many algorithms have been developed since the invention of the computer to solve such problems. Some of these methods are based on existing mathematical approaches, while others are entirely new. We will go over each technique one by one, but first, let us define the optimization problem.

1.2 OPTIMIZATION PROBLEMS

An optimization problem is a real-world situation in which we aim to maximize or minimize a specific objective function f. An objective function, also known as a fitness function, explains how each input vector X is connected to each output Y by a one-to-one mapping, the input domain, and the goal of optimization.

$D(X)$ defines the domain of input values, and it may be finite or infinite. Similarly, $R(Y)$ is the range of output values. Following keywords are used for explaining the design of optimization algorithms

Search space (input domain): Input search space will be represented by $D(X)$, where n-dimensional vector $X = X_1X_2X_3 \ldots X_n$ represents each solution X.

Fitness function: A mapping connecting each X to Y is defined by a mathematical function.

Criteria: It can be maximized or minimize the fitness value.

***Global optimal solution*:** Any value X_{opt} within $D(X)$ will minimize or maximize the value fitness function.

1.2.1 Classification of Optimization Problems

The definition of the optimization problem used in the previous paragraph is generalized. However, depending on the characteristics of the input domain, objective functions, and availability of definition of a function, we can classify these problems into different categories.

1.2.1.1 Classification Based on the Number of Points in the Search Space

Depending on the number of solutions in the search space, optimization problems can be classified as discrete or continuous.

Discrete or combinatorial optimization problem:
Suppose domain *D(X)* of the optimization problem contains a finite number of solutions and forms an extensive configuration search space. In that case, the problem is known as a discrete or combinatorial optimization problem. Examples of combinatorial optimization problems are the Traveling salesman problem (TSP), bin packing, and others.

a. **TSP:**
Input domain of this problem contains positions of N different cities where the position of each city is represented by coordinate (x, y). The goal of the problem is to identify the shortest possible path that visits each city exactly once.
b. **Bin-packing problem:**
The input domain of this problem is a set of N objects, each with predefined size si. The goal is to fit them into as few bins (each of size B) as possible.
c. **Job-shop scheduling problem:**
The job-shop scheduling problem can be designed in many areas such as cloud computing, parallel computing, and operating systems. This problem aims to find an optimal schedule that specifies what jobs should be done, when they should be done, and with what tools while minimizing the total time until all jobs are completed. This problem's input consists of a set of jobs and tools that can be used to meet those jobs.

Continuous optimization problem:
A continuous optimization problem may have an infinite number of solutions in the search space. The input variables of these problems belong to continuous search space. Any real function in which the input domain consists of variables bounded by real values is a continuous optimization problem. An example of a continuous optimization problem is shown by Equation 1.4.

Example:

Fitness Function:

$$F(x, y, z) = x^2 + y^2 + z^2 \tag{1.4}$$

Input Domain:

$$1 <= x <= 4,\ 2 <= y <= 5,\ 6 <= z <= 9$$

Here you can see that real values bound each input variable; hence, this is an example of a continuous optimization problem.

Further continuous optimization problems can be classified as constrained or unconstrained optimization problems.

a. Unconstrained optimization problems

In unconstrained optimization problems, external constraints are not added to inputs, and all solutions in the search space will be feasible. It means there will be no infeasible solution. The solutions of the search space which participated in search of optimal solution(s) are known as feasible solutions. An example of unconstrained optimization problems is shown by Equation 1.5.

Example of unconstrained optimization problems:

Fitness Function:

$$F(x, y, z) = x^2 + y^2 + z^2 \tag{1.5}$$

Input Domain:

$$1 <= x <= 4,\ 2 <= y <= 5,\ 6 <= z <= 9$$

In this example, any input vector can be supplied, and all search space solutions are feasible.

b. Constrained optimization problems:

In some problems, additional constraints are imposed on the domain of the input values, rendering some solutions infeasible. In such cases, some input domain solutions become infeasible. Equation 1.6 depicts a constrained optimization problem.

Example of constrained optimization problems:

Fitness Function:

$$F(x, y, z) = x^2 + y^2 + z^2 \tag{1.6}$$

Input Domain:

$$1 <= x <= 4,\ 2 <= y <= 5,\ 6 <= z <= 9$$

And it must satisfy the following additional constraints:

$$x + y + z <= 8 \ \ and \ \ y + z = 4$$

For this problem, even though (2, 3, 4) belongs to the domain of the problem but it is not feasible as it does not satisfy the additional constraints. However, (1, 2, 3) is a possible solution.

1.2.1.2 Classification Based on the Number of Objective Functions

Optimization problems can also be classified based on the number of objective functions. Only one objective function must be optimized in single-objective optimization problems, whereas multiple conflicting objective functions must be optimized in multi-objective and many-objective optimization problems.

a. **Single-objective optimization:**
Any optimization problem having only one objective function is called as single-objective optimization problem. In such problems, the goal of the optimization algorithm is to scan the input values and produce an optimal solution as soon as possible. These problems can be easily solved by comparing the objective values of the inputs.

An example of a single-objective optimization problem is shown in Equation 1.7.

$$\text{Maximize } f(x) = sin(x) \qquad \text{where } 1 \le x \le 3 \tag{1.7}$$

b. **Multi-objective optimization problems:**
Multi-objective optimization problems require the optimization of two or more conflicting objective functions. Unlike single-objective optimization, the fitness values of solutions cannot be compared to identify an optimal solution. Here, conflicting means if the value of one function increases, the value of the other function decreases. An example of a multi-objective optimization problem is shown by a combination of Equations 1.8 and 1.9

$$\text{Maximize } f_1(x) = sin(x) \qquad \text{where } 1 \le x \le 3 \tag{1.8}$$

$$\text{Maximize } f_2(x) = cos(x) \qquad \text{where } 1 \le x \le 3 \tag{1.9}$$

In this example, *sin(x)* and *cos(x)* are conflicting objective function. When the same input domain is passed in these functions, one will increase, and the other will decrease.

c. **Many-objective optimization problems:**
Any multi-objective optimization problem with conflicting objective functions greater than four is known as a many-objective optimization problem.

1.2.1.3 Classification Based on the Availability of Objective Functions

When it comes to the availability of mathematical functions, these issues can be divided into two categories. The first category of problems is those in which the function to minimize or maximize is explicitly stated. Such issues are referred to as direct optimization problems in this chapter. In other cases, where the function is not given directly, we must formulate it. To create an objective function for these problems, a thorough understanding of the problem domain is required. These are indirect problems that can be applied to any field of study. These are known as indirect optimization problems. Let us take some examples of direct optimization problems and indirect optimization problems.

a. **Direct optimization problems:**
A mathematical optimization problem, in which information of function and search domain is given in the form of a mathematical function, is known as a direct optimization problem. An example of a direct optimization problem is listed in Equation 1.10.

For example:

$$\text{Maximize } F(x) = x^2/(1+x^3) \quad \text{where } 1 \le x \le 4. \tag{1.10}$$

The given equation is an elementary example of an optimization problem. Here we are interested in finding out the maximum value of function $x^2 / (1 + x^3)$. Therefore, $x^2 / (1 + x^3)$ is the objective function. Since it also defines the fitness of any solution, it is also called the fitness function.

To identify the maximum value of the objective function, we need to check the values of the function on all values of x, which is defined as $1 \le x \le 4$. This input domain is called search space, and for this problem, the search space is [1,4].

b. Indirect optimization problems:
There are many optimization problems where the objective function is not directly mentioned. In such problems, we have to study the problem and find the objective function that can be optimized. In some optimization problems, only the values of input and output pairs are known. Here we discuss one issue and show why this is an optimization problem. Let us discuss the situation, which is given in Table 1.1. In this problem, the output of the function is provided for five values. Even though, initially, it does not seems like an optimization problem but when you check it carefully, you will find that it is an optimization problem. Let us see where optimization is involved. When you check Table 1.1 carefully, you will find that output increases for the first few values and decreases after the median value. To identify the optimal value, which produces the maximum value of the function, we have to apply the optimization algorithm. However, as in this problem, mathematical mapping between input and output is missing; therefore, we must first use a method to establish a relationship between input and output. After fixing the function, we have to use optimization techniques to identify the optimal intake. In addition, optimization techniques can also be used to determine the best function that can define the relationship between input and output.

1.2.1.4 Classification Based on the Number of Global Optimal Solutions

Optimization problems can be classified into two categories, (1) unimodal optimization problems and (2) multimodal optimization problems.

TABLE 1.1
Example of Indirect Optimization Problem

F(x)	x
1	1
4	2
16	3
9	4
5	5

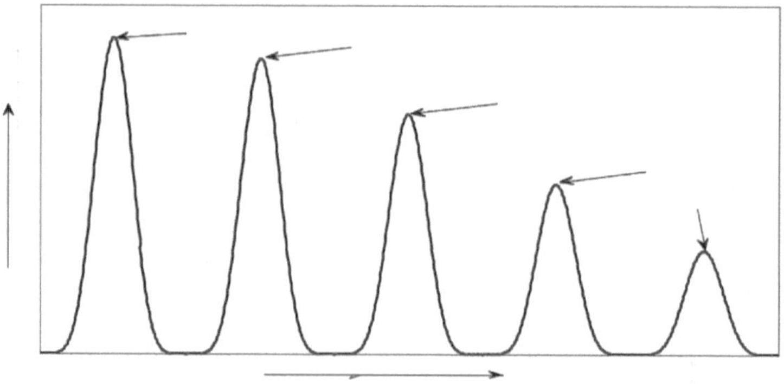

FIGURE 1.2 Multimodal problem with more than one global optimal solution.

Unimodal problems:
Unimodal problems are those problems in which there will be only one optimal solution. So, for designing a randomized algorithm, only two operators are sufficient—one selects a probable area, and the other can perform exploitation.

Multimodal problems:
An optimization problem having multiple local and global optimal solutions is called a multimodal optimization problem. For example, the problem shown in Figure 1.2 represents a multimodal optimization problem. It contains several local and global optimum solutions. In such problems, only selection and exploitation will not be enough. It may produce a local optimum solution. For such problems, selection and exploitation must be combined with exploration.

1.3 METHODS FOR SOLVING OPTIMIZATION PROBLEMS

Let us discuss how we can solve such problems using a computer and what methods will be efficient in solving this problem using a computer.

Many algorithms can be designed to solve these problems. Some of them were described using well-known mathematical approaches in computer programs, while others were created by defining new techniques for searching for the optimal solution. What should we do? Which method will be efficient in solving optimization problems? Let us have a look at both approaches and then decide which one will be more efficient.

1.3.1 Mathematical Methods

In the beginning, gradient-based methods, derivation-free methods, Simplex methods, and a variety of other mathematical techniques were used to solve optimization problems. Although these approaches appear to be very appealing and straightforward, they cause many issues when implementing the program. Due to these implementation issues, these approaches are less popular. Moreover, these techniques are situation-specific and cannot be used as a generalized technique.

For example, the gradient-based method optimizes functions by comparing the gradient values at different points. Unfortunately, writing a program for finding out the gradient value at a given solution is very typical. If somehow, we could implement it, then evaluation of the value of gradient is also costly. Similar kind of problems exists with other mathematical techniques.

1.3.2 Programming Methods

Other methods include the traditional way of solving optimization problems using the computer. For solving any problem using a computer, the programmer has to define an algorithm. An algorithm represents a step-by-step approach for solving a problem by a computer. An algorithm may be deterministic or randomized. Let us discuss what type of algorithms that will work with optimization problems. We explain it with the definition of algorithms and see whether these algorithms can be used for solving the optimization problem or not.

1.3.2.1 Deterministic Algorithm

A deterministic algorithm takes a finite amount of time to produce an output of the problem. Generally, the problem with limited input values is a good candidate for deterministic algorithms.

Example: Sorting problems requires a finite number of inputs to be arranged in proper order. So they can be easily handled by writing deterministic algorithms.

Can we use a deterministic algorithm for solving optimization problems?

No, it will not work for optimization problems. The reason is the number of input values which are very large in optimization problems. For example, in combinatorial problems, input values are enormous and infinite in continuous problems. Due to limited resources and time, applying the deterministic algorithm in such problems is impossible, and a randomized algorithm provides a better solution to optimization problems.

1.3.2.1.1 Designing Deterministic Algorithm for Solving Optimization Problems

Although in optimization problems with an ample input search space, it is quite impossible to write a deterministic algorithm that can produce an accurate global optimal solution, they can be applied if the size of search space is to a reasonable limit. On a smaller version of these problems, we can apply for deterministic program. To create a smaller version of optimization problems, we have to reduce the search space by selecting a fixed number of solutions from the search space. However, before using such kind of division, we have to check the time and resource requirements for the problem. If we have sufficient time, we can select more input solutions in the search space. However, if time is limited, then we have to choose fewer solutions. A strategy for choosing a finite number of solutions from the entire search space can be defined to implement the selection of input solutions. In the following paragraph, we go over these techniques. Sometimes even after reducing the search space, we cannot evaluate all solutions. Due to resource constraints, we have to choose only a subset of solutions to identify the best solution. Remember, only a few comparisons can be made in deterministic algorithms.

Representing solutions of optimization problems and calculating fitness function: A finite number of solutions from the search space can be chosen using a variety of methods. The technique used to divide the search space into a fixed number of solutions is determined by the representation of the search space's solutions. Before using a computer to solve an optimization problem, we must first determine which model will best represent the solutions in the problem's search space. Although several methods can describe optimization problem solutions, in most cases, solutions are characterized as either a binary string or a real-parameter vector. In such problems, we have to divide the search space into 2^n number of solutions where n defines the number of bits in the binary representation. In other issues, where we can directly work with actual values, we can choose real-valued representation. These methods of representing the solutions of optimization problems define the type of encoding used for the problem. If a binary string represents a solution, then the encoding is called binary encoding. If a real-parameter method is used, then the encoding is called real-parameter encoding. After dividing the input search space into a fixed number of solutions, the fitness of each solution is calculated by checking the value of the objective function. Finally, these solutions will be compared based on their fitness values, and the best one will be treated as an optimal solution.

These solutions will be compared based on fitness values, and the best one will be considered the optimal solution.

In some cases, information about the search space and fitness function may not be available directly in numerical form; instead, it may be available in symbolic form, which cannot be processed instantly. Before writing a program, the solutions and fitness functions in such problems must be represented numerically. This can be implemented by an experienced programmer who can use domain knowledge related to the problem to define the fitness function in mathematical form. After determining the function, the solutions to the problems can be represented using binary or real-parameter encoding. Furthermore, these representations are converted into the correct form in order to calculate the fitness function that will guide the search. The type of encoding has a significant impact on the design of optimization algorithms. It may alter the method used to implement operators, affecting the algorithm's performance. Let us discuss in detail binary encoding and real-parameter encoding.

a. **Binary encoding:**

   Binary encoding is used in those problems where a binary string easily represents each solution. Such kind of encoding is mainly used with discrete optimization problems. We can also use binary encoding to solve problems with real-valued inputs. However, it will not be efficient to use binary encoding for solving real-valued problems. There are two reasons behind it. The first reason is the additional efforts needed to convert binary values into actual values required in real-valued problems. Each solution's fitness can be calculated only when the solution is represented in real values in such problems. The second reason is related to the coverage of the whole search space. Using binary encoding, we cannot cover the entire search space in real-valued problems. We will not be able to generate every solution in the search space. The binary encoding technique can represent solutions both in the deterministic and randomized versions of the program. Although nearly all optimization algorithms can be modified to solve problems requiring binary encoding, only a few randomized algorithms are specially designed for handling discrete optimization problems. A simple genetic algorithm is designed explicitly for handling discrete optimization problems. Some other popular optimization algorithms that use binary encoding for solving optimization problems are the Environmental Adaptation Method (EAM), Improved Environmental Adaptation Method (IEAM), binary particle swarm optimization algorithm (BPSO), and binary grey wolf optimization algorithm (BGWO).

Now we discuss some example problems where the binary encoding technique can be used to represent the solutions.

These steps are required to solve a real-valued program by binary encoding.

First, the programmer will decide how many solutions he has to divide the search space. Since a binary value should represent each solution; therefore, the search space is divided into 2^n solutions. After dividing the search space, each solution can be represented by n number of bits. Let us check this example problem to understand how binary encoding is used.

Example problem: Let us take the same example discussed earlier and see how binary encoding can be used to solve this problem.

$$\text{Maximize } F(x) = \frac{x^2}{1+x^3} \qquad \text{where } 1 \le x \le 4$$

To use binary encoding, the solution must be represented by a binary string. This can be done by dividing the search space into 2^n solutions, where n will define the binary string size used to represent each solution. For example, if we want to express each solution by a 4-bit binary number, we must divide our search space into 16 solutions. After dividing the search space into 16 solutions, the first solution can be represented by 0000, and the final solution will be represented by 1111.

Question 1: How many bits will represent each solution if the search space of a given problem is divided into 1024 solutions?

$$\text{Maximize } F(x) = Sin\, x, \quad \text{where } 0 \le x \le 3$$

Answer: In this example, we want to divide the whole search space into 1024 solutions.

So $2^n = 1024$, which will give n as 10.

The number of bits required to represent each solution will be **10.**

Question 2: How many solutions should the search space be divided to represent each solution by 8 bits for the given problem?

$$\text{Minimize } F(x) = Sin\, x\,, \text{ where } 0 \le x \le 3$$

Since the size of the binary string will be 8,

the total number of solutions $\mathbf{2^8 = 256}$.

Question 3: How will you represent the 10th and 256th solutions if the given problem's search space is divided into 1024 solutions.

$$\text{Minimize } F(x) = Sin\,x, \text{ where } 0 \le x \le 3$$

In this example, we want to divide the whole search space into 1024 solutions.

So 2^n =1024, which will give n as 10.

The binary representation used for representing the **10th solution: 0000001001 (decimal number 9).**

For **256th solution = 0011111111 (number 255).**

Converting binary values to actual values in real-valued problems:
In real-valued problems, binary encoding has no connection with the objective function. However, the objective value of each solution is required to guide the optimization algorithm in the proper direction. Therefore, evaluating each solution's fitness is essential for such problems and can be calculated by converting each solution into the actual value. Let us take an example and see how this conversion can be done. Let us suppose that in a given problem, the search space is represented by the following equation: $X_l \mathrel{<=} X \mathrel{<=} X_u$, where X_u is the upper bound of the search space and X_l is the lower bound. Let us suppose that this search space is divided into 2^n of solutions; then, the following equation can convert any binary solution into the corresponding real value, where X_{actual} represents the actual value of the solution.

$$X_{actual} = X_l + \left(\frac{X_u - X_l}{2^n - 1}\right) \times (\delta), \tag{1.11}$$

where δ is the decimal value of the binary string used for representing the solution.

Let us take an example.

Example problem: What will be the actual value of the 10th and 256th solutions if the search space of the problem given below is divided into 1024 solutions.

$$\text{Minimize } F(x) = Sin\,x \text{ , where } 0 \le x \le 3$$

First, we need to write a binary representation of 10th solution. It will be 0000000101.

Upper bound X_u for this problem will be 3, and lower bound X_l will be 0.

The decimal value of 0000000101 will be 9.2^n will be 1024. So the actual value of the 10th solution will be

$X_{actual} = 0 + ((3-0)/(1024 - 1)) * 9 = 0 + (3/1023) * 9 = 0 + 27/1023 = 0 + .02639$
$= \mathbf{0.02639}$

Similarly for 256th solution,

$X_{actual} = 0 + ((3 - 0)/(1024 - 1)) * 255 = 0 + (3/1023) * 255 = 0 + 765/1023 = 0 + .7478$
$= \mathbf{0.7478}$

b. Real-parameter encoding:
In some problems, the variables are bounded by real boundaries. In these problems, each solution can be represented by real values. This encoding technique is known as real-parameter encoding. This encoding technique is generally used in handling continuous optimization problems.

Let us take an example and see how real-parameter values can represent a solution.

Example: In the given problem,

$$\text{Minimize } F(x) = Sin\, x, \text{ where } 0 \le x \le 3,$$

how will you represent the solution by real-parameter encoding?

In a given problem, any real value can represent a solution in a given search space.

For example, x = 1.5 represents one solution represented by real-parameter encoding.

These encoding techniques can be used with both deterministic and randomized algorithms. Let us take some examples to see how these techniques can be used with a deterministic algorithm.

Example problem and application of deterministic algorithm:

Let us see an example of a deterministic algorithm that can be used to solve an optimization problem. A deterministic algorithm for solving the problem Maximize $F(x) = x^2 / (1 + x^3)$, where $1 \le x \le 4$, will divide search space into a finite number of solutions. Evaluate these solutions by checking the fitness value and produce the best solution. While selecting solutions from the search space, we have to choose solutions from the entire search space.

Let us consider four real-valued points in the search space and rewrite the problem again

$$\text{Maximize } F(x) = x^2 / (1 + x^3), \text{ where the solutions are } \{1.0, 2.0, 3.0, 4.0\}.$$

You can see this in this smaller version. We have taken only four solutions. These solutions cover the entire search space and are at the same distance from each other.

So, for each value of *x*, we can identify the value of *F*(*x*), which will be {1/2, 1/9, 1/28, 1/65}.

Which one is the maximum?

1/2 will be the maximum value of the fitness function, and one will be the best solution.

Will it give you the optimal solution to the original problem?

The answer is no.

We cannot use this solution as it was obtained by comparing a significantly fewer number of solutions.

However, if we increase the number of solutions in the search space, we can obtain a more precise solution.

It has been discussed that by dividing the search space into more solutions, we can improve the precision of the optimal solution. However, if we increase the number of solutions, we will not apply a deterministic algorithm. Due to limited resources and time constraints, it is impossible to choose and evaluate many solutions. Let us take an example to understand this problem and see what other methods can be used to provide reasonable approximate solutions.

Let us suppose we divide our search space into 2^{40} solutions to retrieve the optimal global solution to the problem.

Can you assume what will be the complexity of the algorithm in that case?

Complexity will be calculated by adding the evaluations required in these three steps.

- We need to write a mathematical function to calculate all 2^{40} solutions, which will take 2^{40} evaluations.
- After calculating these solutions, we need to calculate the fitness function's value, which will again take 2^{40} evaluations.
- Finally, we have to write an algorithm to identify the maximum or minimum value of the solutions. This algorithm will compare all possible inputs and will take nearly 2^{40} evaluations.

So the time complexity of such an algorithm will be $3 * 2^{40}$. Since this algorithm's time complexity involves exponential terms, it will take time to produce an output of the program.

Even after comparing the 240 solutions, the output of the algorithm cannot be used as an optimal global solution for our basic problem.

We can get a more precise result if we take a larger number of solutions in the search space. However, if we increase the total number of solutions, we will not get output in a reasonable time. Moreover, in such problems, deterministic algorithms may not be implemented due to a lack of resources.

Can we still design an algorithm that produces a good result in less time?

Yes, we can produce good results in less time by using a randomized algorithm.

Can we reduce the time complexity of the problem?

Yes, the time complexity can be reduced if we apply a randomized algorithm in place of a deterministic algorithm.

1.3.2.2 Randomized Algorithms

Many randomization algorithms have been developed to solve optimization problems. Randomized algorithms can produce excellent output in significantly less time.

Randomized algorithms provide efficient solutions to many Non-deterministic Polynomial-time (NP)-Hard problems, including optimization problems. We have already discussed that a nearly optimal solution can be obtained in optimization problems when more solutions are in the search space. However, it also increases the complexity of the problem. Applying the deterministic algorithm becomes impractical in such complex problems either due to the unavailability of resources or the time required to produce the output. For solving such problems, we can design a randomized algorithm. Randomized algorithms do not evaluate all solutions; instead, they will focus on those areas where the probability of finding an optimal solution will be very high. The selection of the probable regions is made with the help of the heuristic function. After selection, exploration and exploitation techniques are combined to search these potential areas to identify optimal global solutions. The knowledge used to identify highly likely areas is formulated in a heuristic function by the programmer. Generally, this knowledge is inherited from some natural phenomena; therefore, these randomized algorithms are also nature-inspired.

Let us take an example to understand the difference between a deterministic algorithm and a randomized algorithm. Moreover, this example will also help the programmer show how problem knowledge can be converted into a heuristic function to design the randomized algorithm.

Example problem: Let us suppose in a class there are 100 students. We want to design a strategy by which we can tell the class's topper as soon as possible.

Defining the solution of a problem using a deterministic algorithm:
In the deterministic version, starting from the first student, we ask the Cumulative Performance Index (CPI), or percentage, of each student, and the student with maximum CPI will be the top of the class. Now we check the complexity of the algorithm. The time and cost associated with the algorithm can be estimated using the number of evaluations required to identify the output. Any reduction in the number of assessments will improve the time complexity of the algorithm. In this algorithm, a total of 100 evaluations will be required to check the CPI of the students, and nearly 100 comparisons will be necessary to identify the maximum CPI. So in total, 200 evaluations will be required to determine the best solution.

1. Will it give you the exact solution?
   Yes
2. Can we design another algorithm, which can produce nearly similar results in less time? If yes, how?
   Yes, we can design such an algorithm, and knowledge related to the problem will reduce the evaluations.

Defining the solution of problem using a randomized algorithm:
In the deterministic version, we have compared the CPI of every solution as it can be seen that in this problem if we compare only reasonable solutions in place of comparing all solutions, the answer will be the same. The identification of a good solution will be dependent on an intelligent strategy. In this example, the teacher will design a plan for reducing comparisons and ask the student to raise their hand if their CPI

is greater than 8.0. In this way, the teacher will evaluate only those students whose CPI is more than 8.0. This query (Intelligent Strategy) decreases the total number of evaluations.

How many evaluations will be required to identify the optimal solution?

Only those solutions will be evaluated which satisfy the intelligent query. Let us suppose that if only eight students are there who have more than 8 CPI. So the total number of evaluations will be limited to 16, which is significantly less than 200. It can also be seen that this method will dramatically reduce the number of assessments.

Why will this solution create a randomized algorithm?

From the preceding example, it is clear that a randomized algorithm will select solutions from random positions.

Will it give you the exact solution?

Not guaranteed, it will give you the near to best solution, but we will not be sure whether this solution will be the best one since the solutions are identified with the help of a query that has been broadcasted to the class. If we suppose the best student is doing some other work, he will not be included in the comparisons. In that case, the best solution will not be obtained. However, the obtained solution will be very close to the optimal one.

From the preceding example, it can be seen that a randomized algorithm with an intelligent strategy can solve any problem in significantly less time. Now we discuss how randomized algorithms can be used to solve optimization problems. How can a smart strategy be designed for optimization problems?

1.3.2.2.1 Designing Randomized Algorithms to Solve Optimization Problems

As we have already discussed, deterministic algorithms can be used to solve a smaller version of the optimization problem. However, these deterministic algorithms will become unsuitable for solving complex optimization problems in which the search space is divided into several points. These solutions are to be evaluated to identify optimal global solutions. In such situations, a randomized algorithm may provide a quick and efficient solution in very little time. Although we follow the same procedure, we have used to implement deterministic algorithms, but it will be modified.

For example, in a randomized algorithm also, we divide our search space into a fixed number of solutions. However, unlike deterministic algorithms, we do not evaluate all these solutions. In place of evaluating all solutions, we start with some random solutions. We consider these solutions to identify which solutions belong to promising regions and which solutions belong to deficient areas. The solution that belongs to bad sites can be discarded as they can never produce an optimal solution. We must exploit the regions around good solutions in a hope that the area will contain more suitable solutions, resulting in a global optimal solution, as good solutions will point to other good solutions. Although exploitation can yield good results, it is insufficient for locating optimal global solutions in particular issues. Exploration should be used in conjunction with exploitation to produce a global optimal solution. So to solve any optimization problem, functions for performing selection, exploitation, and exploration are implemented in randomized algorithms. Generally, a natural phenomenon is used to implement all these techniques. Therefore, these algorithms are also called nature-inspired algorithms.

In nature-inspired algorithms, a natural phenomenon is used for guiding the search toward an optimal solution. The step-by-step process for defining nature-inspired algorithms is shown below.

In a nature-inspired algorithm, we follow these steps to solve an optimization problem.

Step 1: Generating input population: In a nature-inspired algorithm, we start with some randomly selected solutions that are uniformly distributed in the entire search space. These solutions are called the initial population for the algorithm.

Step 2: Create a heuristic function by incorporating natural phenomena: Generally, these optimization algorithms are implemented by simulating a natural process in a computer program. The central part of these biological processes lies in a natural phenomenon that is used to produce new solutions for a given input population. It acts like a heuristic function that guides input solutions in the correct direction.

Step 3: Produce output population: The input population is processed using the heuristic function, producing a new set of solutions. These new solutions will define the output population.

Step 4: Repetition: The process defined in steps 1 to 3 is collectively called iteration. It is repeated until either the desired fitness solution is retrieved or the repetitions reach their maximum count. During the repetition, the previous iteration's output is passed as an input to the next iteration.

Depending on the type of problem we are solving. We have to make some changes in the implementation of exploitation and exploration methods. Before designing a randomized algorithm for solving an optimization problem, we should check the type of optimization problem. Based on the kind of optimization problem, the proportion of exploitation and exploration can be increased or decreased in a randomized algorithm. For example, in unimodal problems, less exploration is required. Let us see the type of optimization problems and their effect on designing algorithms.

Example problem and application of randomized algorithm:
Let us take an example to explain how a randomized algorithm works. Let us suppose we want to identify the optimal solution to this problem.

$$\text{Maximize } F(x) = Sin\, x, \quad \text{where } -10 \le x \le 10$$

For understanding how a randomized algorithm works, let us assume that Figure 1.3 represents the function's graph. From Figure 1.3, we can see that this is a multimodal optimization problem. Since all optimal solutions have the same objective value, we want to capture only one global optimal solution. Similar to the discussed examples so far, we will first divide the search space into as many solutions as we want. Let us assume that we have divided our search space into 2^{40} solutions. However, to reduce the time complexity of a randomized algorithm, we will not evaluate all these solutions. We will compare only those solutions that might produce the optimal global solution.

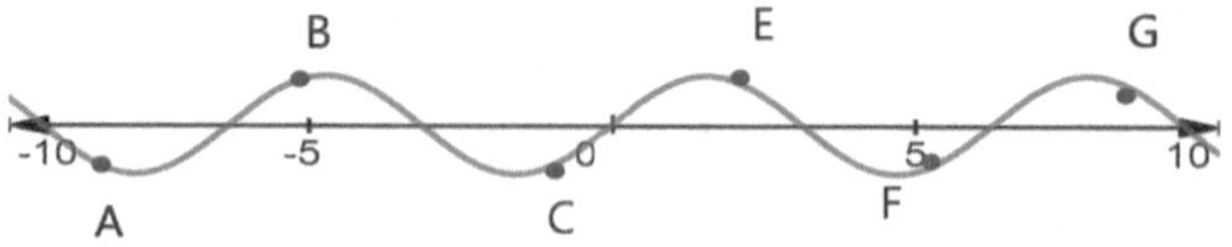

FIGURE 1.3 Graph of sin(X).

After dividing the search space, we will randomly select some solutions from the entire search space. These solutions will define the initial population for the randomized algorithm. In this example, we have taken six solutions as initial solutions. After fixing the initial population, we will apply selection and identify those solutions that can produce optimal solutions. Selection will select all those solutions which belong to suitable regions. After checking the fitness value, it can be seen that solutions B, E, and G belong to promising areas. But the only selection is not enough for producing optimal global solutions. After selection, the area around selected solutions is exploited to produce other good solutions. In addition to exploitation, exploration is also done to check other probable regions. This process is repeated a fixed number of times to capture optimal global solutions.

How to design a randomized algorithm:
To create a randomized algorithm, we have to start with some random solutions selected from the entire search space. These initial solutions will define the initial population for a randomized algorithm. Furthermore, if we accurately guide these solutions in the appropriate direction, they can create more suitable solutions. In randomized algorithms, these initial solutions are directed toward optimal global solutions by some natural phenomena.

1.3.2.2.2 Designing and Establishing New Randomized Algorithms

Many nature-inspired algorithms have been developed by mapping various natural processes. For example, the evolutionary process is used for creating evolutionary algorithms. Swarm intelligence methods are implemented in designing swarm intelligence algorithms. Similarly, other natural phenomena have been used to develop different algorithms. Although several algorithms exist to solve optimization problems, a user may not learn all these algorithms. He wants to use that algorithm that gives the best results in all optimization problems. According to a famous theorem, known as the no free lunch theorem, no nature-inspired algorithm can produce the best results in all optimization problems.

For checking the performance of a nature-inspired algorithm, scientists have created some benchmark functions. These benchmark functions are designed based on various types of optimization problems available in real life. Definitely, the category of all optimization problems is still unknown, so people are developing new benchmark functions that can better represent some problems. These benchmark functions are used for checking the performance of optimization algorithms. For these benchmark functions, the optimal value of the function is already given. The difference between the actual value and the algorithm's optimal value is recorded in each iteration to verify the convergence rate of an algorithm. Sometimes the number of fitness evaluations is counted to judge the performance of an algorithm. A fitness evaluation is measured when the fitness of the solution is calculated.

There are two significant problems with existing optimization algorithms. Some optimization algorithms suffer from low convergence rates, and others may stick to optimal local solutions. For removing these problems, scientists are either creating new nature-inspired algorithms or modifying the existing ones.

Scientists are searching for a new process and creating the computer program for it for developing new algorithms. However, two methods can be used for developing the updated version of existing algorithms. In the first method, parameter tuning is used to improve the performance. In the second method, operators of existing algorithms are updated to enhance the performance.

1.3.2.2.2.1 Framework for Checking the Performance of New Optimization Algorithms After creating a new or updated algorithm, we have to test its performance against existing algorithms. There are two ways to check the performance of any optimization algorithm. Either build programs in any language (C, C++, Java, Python, and MATLAB®) to evaluate and test against the standard benchmarks (available on the IEEE-Congress on Evolutionary Computation (CEC) conference website) or frame programs and run them on popular frameworks like COCO BBOB. Several benchmark function datasets are also available on the websites of CEC conferences, where you can find the information about these benchmarks and read the instruction regarding the testing of optimization algorithms.

Another method used to test the newly created algorithm is to run this algorithm on the existing framework. Several frameworks in several languages are available for testing the performance of new algorithms. Some frameworks, on the other hand, are popular and are used by many programmers. COCO is one such framework. It is a widely used framework for evaluating the effectiveness of any optimization process. The framework's details can be found on the website:

https://coco.gforge.inria.fr/doku.php?id=start

In addition to this, sometimes researchers have developed popular algorithms (like differential evolution, particle swarm optimization, and grey wolf optimization). They also provide codes of these algorithms on their website. These websites may be helpful for a researcher working in this area.

What are benchmark functions?

Benchmark functions are used to check the performance of optimization algorithms. These functions are complex mathematical functions where the value of the optimal solution is already known. Let us suppose that the optimal value of a benchmark function is X_{opt} and the fitness value f_{opt} of this optimal value is represented by $f(X_{opt})$. This fitness value will either be the minimum or maximum value of the fitness function. For checking the performance of optimization algorithms, we run those algorithms on a predefined set of benchmark functions. The performance of algorithms can be compared either based on fitness value produced after a fixed number of iterations or by checking the number of iterations for reaching a solution that satisfies the targeted value for given target function value, $f_{target} = f_{opt} + \Delta f$. The value Δf is defined by the programmer, and in most cases, it is the order of 10^{-8}.

TABLE 1.2
Algorithm and Number of Iterations

Algorithm	Number of iterations used to reach f_{target}
Algo1	200
Algo2	300
Algo3	400
Algo4	500

Let us take an example and see how this criterion can be used to check the performance of the algorithms.

Let we want to compare four algorithms: Algol, Algo2, Algo3, and Algo4. For all these algorithms, the actual fitness of the optimal solution is 0. Let us terminate the algorithm when any population solution reaches the target value $f_{opt} + \Delta f$ and Δf is 10^{-8}.

After running these algorithms, we have collected the following information, which is listed in Table 1.2.

After checking this table, we can identify that Algol is reaching the target value in minimum iterations, the best algorithm.

Instead of comparing the number of iterations, we can sometimes count the total number of function evaluations. This criterion is superior to counting iterations because it depends on the total number of assessments required to achieve a given optimal value. When you evaluate a solution's fitness value, you calculate a function evaluation. If a person considers a solution twice in one generation, the fitness evaluations will be twice as many iterations.

The benchmark functions used to test the optimization algorithm's performance can be scalable or fixed. The dimension of an optimal solution refers to the number of input variables needed to describe it. Only one dimension is used for fixed benchmark functions. The notion of a scalable benchmark function can be expanded to include variable dimensions. The complexity of an optimization issue rises as the number of dimensions grows. As a result, searching for optimal solutions in greater dimensions has become commonplace. So for high-dimension functions, more fitness evaluations will be required to reach the optimal global solution.

Since nature-inspired algorithms are randomized algorithms, they may sometimes capture optimal global solutions in very few iterations. For a fair comparison, we need to run algorithms at least 30/50 times, and then we have to check the average fitness captured in each iteration.

Scientists have created many benchmark functions. These benchmark functions are intended to test the performance of optimization algorithms on a variety of optimization problems, and each set of benchmark functions represents a different type of problem. For example, a set of benchmark functions can represent unimodal problems, whereas another set of functions can represent multimodal problems.

Other categories are also created in the same way. The optimal solution of the benchmark function was fixed in CEC benchmarks from 2005. These benchmarks are no longer beneficial. People are currently working with shifted functions. It is difficult to capture the optimal global solution in such benchmark functions. In such benchmarks, the position value of the optimal solution is moved to a new position each time your experiment is run.

2 Binary Genetic Algorithms

2.1 INTRODUCTION

Many researchers have designed optimization algorithms by mapping the theory of natural selection and evolution. These optimization algorithms are called evolutionary algorithms. Many popular algorithms, such as genetic algorithms (GAs), differential evolution algorithms, evolutionary strategy, evolutionary programming, and genetic programming, come under evolutionary algorithms. These algorithms implement the process of evolutionary learning for designing optimization algorithms. In general, evolutionary learning with the help of three operators known as selection, crossover, and mutation. Like in GA, differential evolution algorithm, and genetic programming, the learning strategy is implemented by three operators, that is, selection, crossover, and mutation. However, in evolutionary programming and strategy, the learning is implemented by two operators known as selection and modification. In this chapter, we discuss GAs.

2.2 GAS

Prof. John Holland proposed GAs in 1977 (J. H. Holland, 1984). These algorithms are population-based algorithms that refine randomly generated solutions repeatedly until an optimal solution is captured. The conversion from the initial population to the final population is implemented in several steps, also known as iterations. Selection, crossover, and mutation operators convert the input population into the output population in each stage or iteration. Since GA is a population-based algorithm, it processes many solutions in one iteration. These solutions are said to form a generation. For example, initial solutions generated randomly from the search space define the initial or 0th generation solution. These solutions are updated through selection crossover and mutation during the first iteration to produce a new set of solutions. These newly generated solutions belong to 1st generation. After the first iteration, the freshly generated solutions are again passed into selection, crossover, and mutation to create the next set of solutions. The exact process is repeated again and again until the termination criteria are reached. The genetic algorithm program is terminated when either number of iterations comes to the maximum value of iterations or the solution with desired fitness is obtained. Several versions of GA have been created to satisfy the requirements of different problems. These versions are named according to the encoding used for representing the solutions in GA. For example, if the binary representation of the solution is used in implementing GA. It is known as Binary GA. If real-valued encoding is used, then the GA is called real coded GA. In the upcoming sections of this chapter, we discuss the details of these versions.

DOI: 10.1201/9781003313649-2

2.2.1 Some Basic Requirements for Designing Optimization Algorithms

Let us discuss some basic needs of developing optimization algorithms. These requirements are common to all optimization algorithms and should be followed to produce good results in fewer generations.

Let us take one example and see what we want to do with the help of an optimization algorithm. Figure 2.1 shows the graph of an objective function in which we want to identify the maximum value.

You can see the function has global as well as local optimal solutions. An optimization algorithm must exploit all probable areas in the search space to capture the optimal global solution to solve these problems. If exploitation is performed only in specific regions of search space, the optimization algorithm will not charge the optimal global solution. For writing an effective optimization algorithm, exploitation of all probable areas should be done correctly. It means exploring the whole search space is an essential factor in designing the optimization algorithm.

Similarly, the algorithm will not obtain a global optimal solution if the search space is explored correctly, but probable areas are not appropriately exploited. Therefore, for writing a good algorithm, a balance between exploitation and exploration is required. As a result, instead of focusing on the area surrounding good solutions, we always strive to cover the entire search space. However, no technique can tell you whether the whole search space has been explored. As a result, we must define some standard methods that will satisfy us that we have attempted to cover the entire search space. For example, in all nature-inspired algorithms, population initialization is used to select solutions of the whole search space. After generating initial solutions, the fitness of each solution is calculated, and only the best solutions are chosen for exploitation. This selection is made at random to broaden the scope of the search. Once reasonable solutions have been identified, exploitation and exploration are carried out with the help of operators. These operators will make every effort to cover the entire search space and exploit every possible area. These operations are

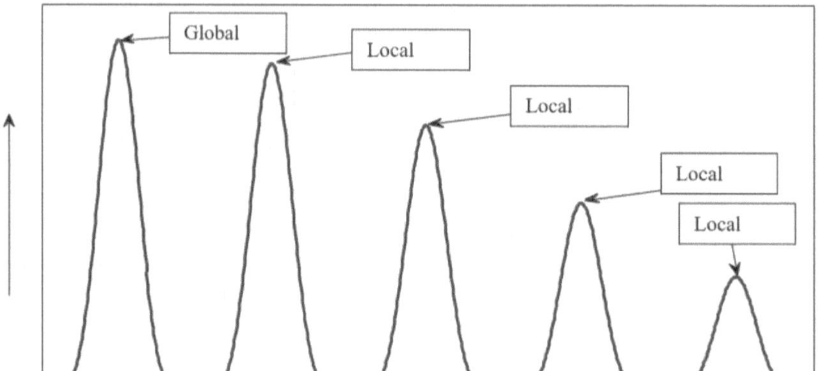

FIGURE 2.1 Multimodal optimization problem.

used to search for solutions iteratively until the best solution is found. Let us take a look at how these techniques are used in GAs.

Alike other nature-inspired algorithms, a GA also generates an initial population from the whole search space. These initial solutions are uniformly distributed in the search space and cover the entire search space. This kind of initialization helps cover the entire search space and avoids the problem of stagnation. After generating the initial population, selection, crossover, and mutation operators are applied to the parent population to produce the following solutions (offspring population). These operators were designed to map the actual process of natural selection and evolution. For example, the selection operator of a GA creates multiple copies of good solutions and discards terrible solutions. The selection operator works as a filter that promotes only fit individuals to survive and reproduce. After selection, fit individuals will go through crossover and mutation and produce offspring whose genes are better than both the parents'. These newly created offspring will again pass as input for selection, crossover, and mutation to create a new set of offspring. This process is implemented again and again until the best solution is obtained.

2.2.2 Solving Optimization Problems Using a GA

Before using GA to solve an optimization problem, we must identify which GA version will be suitable for handling the optimization problem. The two most used versions of GA are binary GA and real-parameter GA. Binary GA is extremely useful in solving discrete optimization problems, whereas real-parameter GA helps solve continuous optimization problems. In binary GA, a binary representation is used to represent the solution of the problem. Similarly, in real-parameter GA, real-valued variables are used for expressing input variables. The method for representing a solution in GA defines the type of encoding. Let us have a look at different types of encoding techniques used for describing the solution.

2.2.3 Encoding Techniques Used in GAs

GAs provide flexibility in using different kinds of encoding for solving problems. Other types of representation can be used to represent solutions while solving a problem through the GA. An encoding technique changes the implementation of selection, crossover, and mutation operators. It means that operators used in binary genetic algorithms will be different from real-parameter algorithms. In addition to binary and real-parameter encoding, other encoding techniques like grey encoding, quantum encoding, and permutation encoding can also design different versions of genetic algorithms. The solution is also referred to as a chromosome in a genetic algorithm encoded to represent the individual's various traits properly. The most widely used encoding is binary encoding, where the solutions are encoded as binary strings. Each bit of the string indicates a specific feature of the solution domain. Depending on the problem, various types of encodings are possible

a. Binary encoding:

In this encoding, the chromosome is represented as a string of 0s and 1s. Let us look at an example of how binary encoding can be used. This type of encoding is extremely effective in solving discrete optimization problems. It can also be used to solve real-valued problems at times. Assume we want to maximize the function *f(x)* in Equation 2.1.

$$Maximize\ F(x) = x^2 / (1 + x^2), \tag{2.1}$$

where the search space is defined as $1 <= x <= 4$.

The problem mentioned in Equation 2.1 is a continuous optimization problem as the search space shown in Equation 2.1 is continuous. However, this problem can also be solved with the help of binary encoding. Let us see how binary encoding is used for solving this problem.

If we divide this search space into 256 points/solutions, each solution can be represented by 8 bits. In Table 2.1, three solutions are selected randomly and are expressed as binary numbers.

A solution represented in binary defines the chromosome in the GA. So, 10011001 is one chromosome. We can also write chromosomes as genotypes. Each chromosome determines the genes of solutions. For example, chromosome 10011001 contains eight genes. If we start with the first digit first gene will be 1, the second gene is 0, the third gene is 0, and similarly, we can list other genes.

In some problems, binary coding will not be effective. Permutation-based encoding is another encoding used to solve the problems.

b. Permutation encoding:

Sometimes we use permutation-based encoding for solving problems. This type of encoding is used primarily in solving graph-related problems. Examples of permutation encoding are shown in Table 2.2. In this encoding, the chromosomes are represented as strings of numbers.

TABLE 2.1
Binary Encoding

CHROMOSOME 1	10011001
CHROMOSOME 2	10100110
CHROMOSOME 3	11101010

TABLE 2.2
Permutation Encoding

CHROMOSOME 1	3 9 8 2 5 1 6 7 4
CHROMOSOME 2	5 7 2 6 8 1 9 4 3
CHROMOSOME 3	7 2 5 1 3 6 9 4 8

TABLE 2.3
Value Encoding

CHROMOSOME 1	A09:231341
CHROMOSOME 2	NDS_12354I
CHROMOSOME 3	2+1345=311

c. **Value encoding:**
Each chromosome is represented by a string that may contain anything from a number to characters in this encoding. Some examples of value encoding are shown in Table 2.3

d. **Real-parameter encoding:**
The most commonly used encoding technique is real-parameter encoding. In this encoding, a solution is represented by real values.

Let us see one example:
Let us suppose that we want to maximize a three-variable function defined as

$$\text{Maximize } F(x_1, x_2, x_3) = x_1^2 + x_2^2 + x_3^2, \tag{2.2}$$

$$\text{where, } 1 \le x_1 \le 5,\ 2 \le x_2 \le 6,\ 1 \le x_3 \le 7. \tag{3.3}$$

Equation 2.2 defines the objective function, and Equation 2.3 is used for defining the search space. Using real parameter encoding, each solution can be represented as a real-parameter vector, where each variable is generated randomly within the given boundaries of the variables. Table 2.4 contains some real parameter solutions that are represented by real-parameter encoding.

Example:

TABLE 2.4
Real-Parameter Encoding

CHROMOSOME 1	(2.5, 0.5, 0.6)
CHROMOSOME 2	(3.0, 4.0, 1.0)
CHROMOSOME 3	(2.3, 3.5, 2.4)

2.3 DETAILED IMPLEMENTATION OF BINARY GAs

In a binary genetic algorithm, search space is divided into 2^n number of solutions. This division is made to represent each solution as a binary number. The binary encoding used for expressing the solution also defines the genotype of the solution. Sometimes the word *chromosome* is also used for this representation. Each chromosome combines several genes which determine the individual trait of an individual. Each solution has its own set of unique characteristics. Each bit in the chromosome represents one gene of the solution. The GA process starts with selecting an initial population by picking random binary solutions from the search space. The population size is dependent on the complexity of the problem and is defined by the programmer. In most cases, the size of the initial population is taken as 40 solutions. After creating the initial population selection operator is used to mark good solutions in the population.

Different methods are used to deal with unimodal and multimodal problems. In the first chapter, we saw that the unimodal problem has only one global optimal solution. As a result, the selection criteria will be based on the solution's fitness in such cases. Selection operators generate multiple copies of solutions with higher fitness and discard solutions with a lower fitness in unimodal problems. Following selection, new solutions around the selected solutions are exploited to identify the best global solution.

However, in the multimodal problem, this method of selection will not work. If we select solutions according to fitness only and then exploit these areas, we may be in a local optimal solution. To capture optimal global solutions in multimodal problems, the selection operator must select those solutions representing different areas of the search space. In such problems, the sharing-fitness method will work best.

In unimodal problems, the selection operator creates multiple copies of good solutions where the solution's fitness measures the goodness of a solution. A fitness function evaluates the fitness of an individual. The lesser the fitness of an individual, the less likely is he to be selected for reproduction. However, the size of the population remains constant after selection. The selection operator will not produce any new solution; only crossover and mutation generate new solutions. The newly developed solutions retain some properties of their parent by storing one or more genes from their parents. Recombination of the parent genes occurs during reproduction, and their offspring inherits both parents' traits. A mutation in the genes of the offspring might occur as well. The fitness of the offspring is evaluated with the next generation to see if they are fit to be a new parent. This cycle continues until we achieve the desired result or reach the maximum number of iterations (generations). Another name used for the binary genetic algorithm is a simple genetic algorithm (SGA). The pseudocode for SGA follows:

```
Function SGA()
{
    Initialize population;
    Calculate fitness function;
    While(fitness value != termination criteria)
    {
        Selection;
        Crossover;
        Mutation;
        Calculate fitness function;
    }
}
```

The step-by-step explanation of the pseudocode follows:

Step 1: Initialize population: Generate the initial population POP (binary strings) of N individuals. POP will work as a parent population and will produce offspring.

Step 2: Calculate fitness function: A fitness function calculates the fitness values of solutions in the population.

Step 3: Selection: The selection operator removes the unfit solutions that are not allowed to reproduce. If an individual is deemed fit by the fitness function, it remains in population POP. Suitable solutions will create multiple copies. There will be no change in the size of the population. Re-create POP after selection.

Step 4: Crossover and Mutation: Check the probability of crossover and mutation. The genetic operators (crossover, mutation) are applied to create offspring OOP from the parent population POP. The size of the new population created after crossover and mutation will remain the same as that of the old population.

Step 5: While (fitness value!= termination criteria): The condition in the while loop is checked, and if not true, then the children population OOP becomes the new parent population POP and forms a loop from steps 2 through 4 until the maximum iterations are reached or the optimum is achieved.

In one iteration of binary GA, selection, crossover, and mutation are applied in the parent population. Many selections, crossover, and mutation methods were proposed to design binary GA. Let us see in detail the working of those operators, and let us try to find out which selection crossover and mutation operator is used in which condition.

2.3.1 Types of Selection Operators

All solutions of the parent population will not participate in the reproduction process. Only a portion of the population can reproduce and continue their generation, and the selection operator identifies these solutions. In unimodal problems, the fitness function determines the fitness value of all the individuals, and the individuals with the best traits get the highest fitness value. A higher fitness value dramatically increases an individual's chance to get selected as, according to the theory of evolution, only the fittest should survive and reproduce for a species to exist.

The various kinds of selection operators used in GAs are discussed in the following subsections.

2.3.1.1 Fitness Proportionate Selection/Roulette Wheel Selection

The **roulette wheel** selection or proportionate selection operator selects solutions based on how well they contribute to the overall fitness of all solutions. In this selection, the total fitness of all solutions is calculated by adding the fitness of all solutions. Following the total fitness calculation, total fitness is normalized between 0 and 1, and a wheel representing the total fitness of all solutions (beginning at 0 and ending at 1) is drawn. This wheel is then divided into different segments according to the contribution of each solution in determining overall fitness. The proportion of each solution in making a fitness wheel is determined by dividing the fitness of each solution by total fitness. Once the proportions of all solutions are identified, the contribution of each solution i in making the wheel is defined by their cumulative

TABLE 2.5
Roulette Wheel (RW) Selection

Chrom #	Binary Solution	Actual Solution	Fitness	% of RW	Cumulative Probability	Solution Selected
1	0101	1	4	8.44	0–.0844	2
2	1010	2	9	19.0	.08.44–.2745	0
3	1111	3	16	33.7	.2745–.6115	1
4	0011	.6	1.96	4.1	.6115–.6525	1
5	0110	1.2	4.84	10.2	.6525–.7545	1
6	1100	2.4	11.56	24.4	.7545–1	1
			47.36	100		6

probabilities <Cli,Cui>. Since the proportion of solution depends on the solution's fitness, the fittest individuals will cover the largest share of the roulette wheel, and the weakest have the least. For selecting a solution using roulette wheel selection, a random value is generated between 0 to 1. A solution whose cumulative probabilities contain this random number will be selected for participating in mating. The fitter individuals have a large portion, so they are more likely to be selected.

Let us take an example to explain the process of roulette wheel selection. Let us suppose we want to apply roulette wheel selection in a population of six solutions. The fitness function used for optimization is $f(x) = x^2 + 2 * x + 1$. The value of x lies between $0 <= x <= 3$. We have initialized the population, and the following solutions are picked in the initial population. The data is given in Table 2.5. Here you can see the fitness value is directly given in the example. Fitness function will be given to you in some problems, and you have to identify the fitness value.

Following the assessment of fitness, the sum of fitness (total fitness) is computed. In this case, the answer is 47.36. After calculating these values, the contribution of each solution is calculated by dividing the fitness value of each solution by total fitness. According to Table 2.5, the first solution's contribution to defining total fitness is 8.44%. Similarly, the contributions of other solutions are calculated. The proportion of each solution in the fitness wheel can be represented by two cumulative probabilities $<c_1, c_u>$, where c_u will not be included in the solution. Similarly, the cumulative probabilities of other solutions will be calculated by adding the percentage value of each solution. The proportion of solution one in total fitness will be from <0, 0.0844> remember lower value .0844 will not be included in the first solution. Similarly, the proportion of solution 2 will be from <0.0844, 0.2745>. So 0.2745 will not be included in solution 2, solution 3 will be <0.2746–0.6115>, solution 4 will be <0.6116–0.6525>, solution 5 will be <0.6525–0.7545>, and solution 6 will be <0.7545–1>. For selecting a solution, a random value is selected between 0 to 1. Let us suppose the selected value is 0.39. This value comes under solution 3, so solution 3 will be selected. Figure 2.2 shows how the wheel is divided among different solutions.

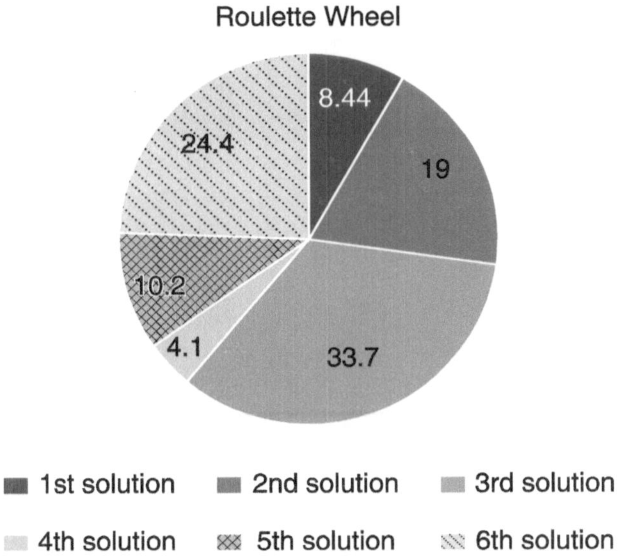

FIGURE 2.2 Roulette wheel selection.

2.3.1.2 Tournament Selection

Tournament selection refers to the process of selecting individuals through the use of tournaments. In this selection, two or more individuals are chosen at random (based on the selection rate) and compete in a tournament; the tournament's winner is chosen for reproduction. The tournament selection procedure is as follows: a random number between 0 and 1 is generated and compared to the fixed selection probability. The fitter individual is chosen if this random value is less than or equal to the set selection probability; otherwise, the weaker individual is chosen. To increase or decrease the selection pressure the selection probability can be changed.

Let us take an example where solutions shown in Figure 2.3 are participating in tournament selection. Let us see what will happen when a tournament is played between two selected solutions if the selection probability is .70 and if the criterion is maximization.

See what will happen when these solutions participate in the selection.

Let us assume that the fitness function is defined by $f(x) = x^2 + 2 * x + 1$, and x is directly represented in the solution. Let us suppose that if the first tournament is played between 22 and 13, the random number generated is .46. Solution 22 will win the tournament. Let us consider for the second tournament, the random number is .85. This will select solution 32. Similarly, solutions selected in other tournaments can be identified.

2.3.1.3 Rank Selection

In rank selection, an individual is selected based on relative fitness among the population. It is better than roulette wheel selection. In roulette wheel selection, the fitter

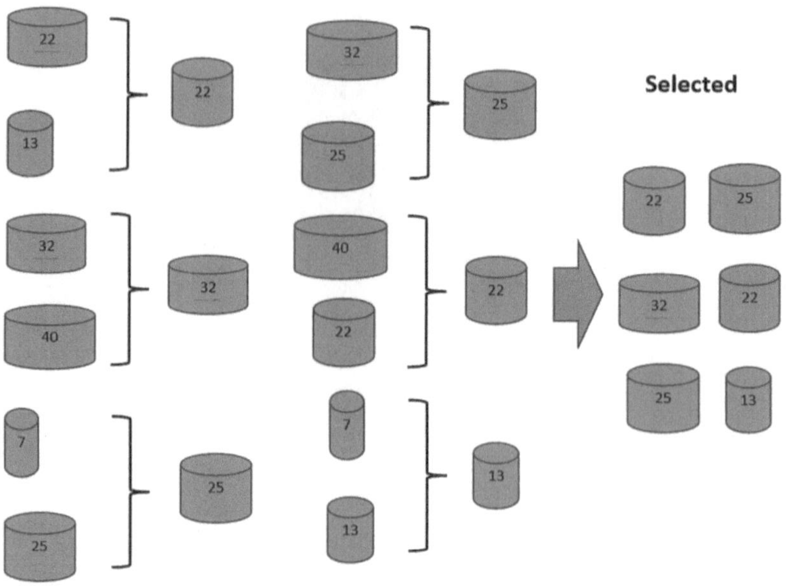

FIGURE 2.3 Tournament selection.

individual will cover more portion of the wheel, so if an individual is very fit, they might take up almost all the wheel, which won't give a chance to the other individuals to get selected. In rank selection, it does not matter if an individual is a hundred times fitter than the others; what matters is the rank they have been given. It not only prevents an early convergence of the algorithm at a local minimum, but it also slows the convergence rate.

Rank selection is a modified version of roulette wheel selection. In this selection, after generating the contribution of each solution. Solutions are sorted according to fitness, and a rank is assigned to each solution. A solution with better fitness is given a higher ranking.

Let us take the same example that we have taken for roulette wheel selection. After the contribution of each individual is identified, they are ranked according to the criteria. In the problem, we want to maximize fitness so a solution with higher fitness will be assigned more fitness. See Table 2.6 to observe how the rank is generated for the solutions.

2.3.1.4 Sharing Fitness Method

In a unimodal problem, as only one global optimal solution exists, only solutions near the global optimal solution will have higher/lower fitness. The solutions that will not be close to global optimal solution will not have good fitness function values. When a fitness-based selection method is applied in such problems, it will promote the selection of solutions that belong to a good region and have higher/lower fitness. Once such solutions are selected, the area around these solutions can be exploited using crossover and mutation operators to produce an optimal solution as early as possible.

TABLE 2.6
Rank-Based Selection

Chrom #	Binary Solution	Actual Solution	Fitness	% of Roulette Wheel	Rank of Solution
1	0101	1	4	8.44	2
2	1010	2	9	19.0	4
3	1111	3	16	33.7	6
4	0011	0.6	1.96	4.1	1
5	0110	1.2	4.84	10.2	3
6	1100	2.4	11.56	24.4	5
			47.36	100	

Although fitness-based selection works effectively in unimodal problems, it cannot solve multimodal problems. Since there might be several local and global optimal solutions in multimodal problems, fitness-based selection may select solutions from some specific area and ignore solutions around the optimal global solution. Once this region is exploited it may misguide the search and can end up in a local optimal solution. So, what should we do in such problems?

In multimodal problems, the requirement is different. Although we want to promote selecting good solutions with higher fitness, we want to avoid selecting multiple solutions from a specific region. We want to design a selection operator that selects solutions with higher fitness and selects them from various areas. Let us take an example and think about how this kind of selection can be implemented in GAs.

For selecting good solutions from multiple regions, the sharing fitness-based selection was developed by David Goldberg. The idea of this selection method is very simple. In this selection method, in place of comparing the actual fitness of the solutions, we first calculate the sharing fitness of each solution and then apply the selection. The sharing fitness of each solution is calculated by dividing each individual's fitness with its niche count. The value of the niche count of a solution defines whether the solution is sharing the region with other solutions or not. If the solution is surrounded by other solutions, then the value of niche count will be higher. If the region is not crowded, then the niche count will be less. Using this method, we can decrease the fitness of those solutions representing the same region and increase the chance of selection of inferior solutions representing less crowded regions.

For calculating the value of sharing fitness of a solution first its fitness value is calculated. After calculating the fitness, the value of the niche count is calculated. This count is calculated with the help of sharing fitness function, which identifies whether a solution is sharing a region with another or not. The calculation of sharing fitness is done on the basis of threshold distance σ, which the programmer defines. If the Euclidian distance d_{ij} between two solutions will be less than threshold distance σ, then they will be sharing the same region. If it is greater than the threshold, they will not have any sharing effect.

The formulae for calculating sharing distance, niche count, and sharing fitness are shown in Equations 2.4, 2.5, and 2.6, respectively.

$$SH(d_{ij}) = \begin{cases} 1 - \frac{d_{ij}}{\sigma}, & \text{if } d_{ij} \le \sigma \\ 0, & \text{if } d_{ij} > \sigma \end{cases} \tag{2.4}$$

$$nc_i = \sum_{j=1}^{n} SH(d_{ij}) \tag{2.5}$$

$$fi_{sh} = \frac{f_i}{nc_i} \tag{2.6}$$

This will degrade the actual fitness of the solutions if other solutions surround them. It means the solutions that are sharing the same region will have less sharing fitness. To select good solutions from multiple areas, we have to select only some representative solutions from each area. It means that if more than one good solution belongs to the same area, we have to decrease the fitness of each solution. In this way, the chances of selecting those solutions will be higher, representing a different region.

2.3.2 Types of Crossover Operator

Crossover between the chromosomes of the parent solutions produces offspring with mixed traits from both the parents. If there were no crossover, there would be no difference between the child and parent. Crossover is the intermixing of parent chromosomes at random selection points known as crossover points to produce children with few traits from the left side of the crossover point and others from the right side of the crossover point.

Crossover is what makes the solution diverse and explores the solution space. The different types of crossovers are discussed in the following subsections.

2.3.2.1 Single-Point Crossover

A single random crossover point is selected, and the children have traits from the left side of the crossover point from one parent and the right side of the crossover point from the other and vice versa.

Tables 2.7 and 2.8 explain how a single-point crossover can be applied.

Example:

TABLE 2.7
Parent Population before Crossover

PARENT A	**10011|001**
PARENT B	**10100|110**

Note: (where '|' represents the crossover point.)

TABLE 2.8
Children Generated after Crossover

CHILD A	**10011|110**
CHILD B	**10100|001**

2.3.2.2 Two-Point Crossover

In a two-point crossover, instead of one crossover point, two crossover points are selected at random, and the children are produced by intermixing traits from both crossover points. Tables 2.9 and 2.10 explain the process.

Example:

TABLE 2.9
Parent Population before Two-Point Crossover

PARENT A	**100|11|001**
PARENT B	**101|00|110**

('|' represents the crossover point.)

TABLE 2.10
Children after Two-Point Crossover

CHILD A	**100|00|001**
CHILD B	**101|11|110**

Note: ('|' represents the crossover point.)

2.3.2.3 Uniform Crossover

In a uniform crossover, many crossover points are selected at random, and the bits are inherited uniformly by both the children. Tables 2.11 and 2.12 explain how uniform crossover can be used.

Example:

TABLE 2.11
Parent Population before Uniform Crossover

PARENT A	**1|00|1|1|00|1**
PARENT B	**1|01|0|0|11|0**

('|' represents the crossover point.)

TABLE 2.12
Children after Uniform Crossover

CHILD A	**1|01|1|0|00|0**
CHILD B	**1|00|0|1|11|1**

('|' represents the crossover point.)

2.3.2.4 Arithmetic Crossover

In this type of crossover, arithmetic operations are done on the parent chromosomes and the children chromosomes are born because of those arithmetic operations. Tables 2.13 and 2.14 explain the functioning of the arithmetic crossover.

Example:

TABLE 2.13
Parent Population before Arithmetic Crossover

PARENT A	**10011001**
PARENT B	**10100110**

('|' represents the crossover point.)

TABLE 2.14
Children Generated after Arithmetic Crossover

CHILD A(AND)	**10000000**
CHILD B (OR)	**10111111**

2.3.3 Types of Mutation Operator

Although mutation operator can be used both for implementing exploration and exploitation, yet in most cases, it is generally used for producing diverse solution in search space. Two commonly used mutation operators are one-bit mutation and multibit mutation. The number of mutations to be implemented in a particular generation is dependent on the mutation probability used in that generation of a GA. It helps maintain genetic diversity and prevents the algorithm to stop at the local minima. Without mutation, the population may become too similar causing the evolution to either slow down or even completely stop.

Example of mutation operator:

TABLE 2.15
Parent Solution before Mutation

PARENT A	**1010<u>0</u>011**

("_" represents the mutation operator)

TABLE 2.16
Child Generated after Mutation

CHILD A	**10101011**

2.3.3.1 One-Bit Mutation

One-bit mutation is used to create new offspring by flipping one bit of the parent solution. One-bit mutation operator causes the change in one random bit of the parent string to produce offspring.

Example of one-bit mutation:

Parent	**10100011**
Child	**10101011**

2.3.3.2 Multibit Mutation

Multibit mutation is used to create new offspring by flipping more than one bit of the parent solution. Multibit mutation operator causes the change in more than one random bit of the parent string to produce offspring.

Example of multibit mutation:

Parent	**10100011**
Child	**11101011**

2.4 THEORETICAL FOUNDATION OF GAs (SCHEMATA THEORY)

Schemata theory is very helpful in explaining why a GA program is able to capture a global optimal solution. Why does a GA converge to a global optimal solution? The foundation of schemata theory lies in defining the word *schema* and showing how schema evolves when selection, crossover, and mutation are applied in GAs.

A schema is used to define similarities in different strings. It works as a template for defining a set of strings where fixed values are represented by either 0 or 1 and variable bits are represented by *. Let us take one example and see how schema can be represented for three strings.

Let us suppose if strings are 1001001000, 1000010011, and 1001111111.

It is clear that in all these strings, the first three bits, 100, are fixed, and the remaining seven bits may be 0 or 1. Hence, the schema for these three strings can be defined as 100*******. You can see in the schema that fixed bits are represented by 100 and other bits are represented by the do-not-care symbol *. Although this schema is defined with the help of only three strings, but this schema can be used to generate 2^7 substrings.

Similarly, other schemata can be defined which can be used as a template for representing a set of strings.

For example, 00*00*110* is able to generate 8 number of substrings.

For establishing the theoretical foundations of GAs, we will try to figure out how the schema evolves with a GA. What will happen to the schema that represent good solutions? Whether such schema will be able to survive after selection, crossover, and mutation determines if they will be discarded. What will happen to those schemata that represent bad solutions? Will they survive after selection, crossover, and mutation?

There are two important keywords, which are associated with schemata theory. These keywords are the order of schema and the defining the schema's length. Let us see the definition of these keywords.

Order of schema: Order of schema represents the total number of fix bits in a schema. It is represented by symbol o.

For example:

In schema 100***100***,

o(S) = 6 because 6 bits of this schema are fixed.

Defining length of schema: Defining length of schema is determined by subtracting the last position and first position of fixed bits in the schema. It is represented by δ.

For example, 100***000**1.

The defining length of this schema will be 12 – 1 = 11.

Now with the help of order and defining length, we explain how schema evolves:

2.4.1 Effect of Selection on a Schema

The selection operator is used in a GA to create multiple copies of good solutions and to remove bad solutions from the population. Let us take a look at how the selection operator affects schema.

Assume that we have already started the GA population and that a binary string represents each solution. Furthermore, keep in mind that all these strings can be defined by multiple schemata. Each schema will generate a set of substrings that can be combined to form different blocks during the initial population. Let us look at an example to see what we mean.

Let's say the population size in a given problem is 10 and the length of the binary string is also 10. Assume the population is represented by various strings, with $S_i(t)$ being the ith string in the population at time t and $F(S_i(t))$ denoting the fitness of the ith string at time t. Furthermore, suppose that any solution's fitness is proportional to the number of ones in the string and can be measured by counting the total number of 1s in the string.

The fitness value for 1001001000 will be 3. Other solutions' fitness can be determined in the same way. The initial population follows:

1001001000
1000010111
1001001111
0101110001
0101000011
0101100110
0001110001
1111111110
0011110000
0011111111

Let us take one schema S_i and see how it evolves with selection. Let us represent this schema as $S_i(t)$. 100*0***** is one of the schemas that represent some solutions at time t.

The fitness of a schema is defined as $F(S_i(t))$, and it can be calculated by averaging the fitness of all solutions in the population created by the given schema.

The schema 100*0***** is defined by three strings {1001001000, 1000010111, 1001001111}, and the fitness of this schema will be (3 +5 + 6)/3 = 14/3 =4.6

As we know in selection, the selection of a solution depends on the solution's fitness. Let us suppose if F(P(t)) is the summation of fitness of all solutions at time t then the probability of selecting any individual will be $F(S_i(t))/F(P(t))$. The chances of survival of this solution will depend on this probability.

To determine whether a schema will survive in the next generation. We must evaluate the schema's fitness as well as the chance of survival. The chances of survival for schema S_i at time $t + 1$ can be represented by the schema $S_i\ (t + 1)$ will be dependent on the fitness of the schema.

Let us represent the probability of getting schema S_i in new iteration $(t + 1)$ as

$$S_i(T+1) = S_i(T) * Population\,size * \frac{Fitness\,Of\,Schema\,at\,Time\,T}{Fitness\ of\ Population}. \tag{2.7}$$

It can also be represented as

$$S_i(T+1) = S_i(T) * \frac{Fitness\,Of\,Schema\,at\,Time\,T}{Average\,Fitness\ of\ Population}, \tag{2.8}$$

where $average\ fitness\ of\ population = \dfrac{fitness\ of\ population}{population\,size.}$

From the Equation 2.8, it can be deduced that the schema whose fitness is greater than the population's average fitness will increase exponentially in future generations.

And solutions with lower-than-average fitness will be phased out of the next generation. This theory also demonstrates that GA theory encourages the creation of good solutions.

2.4.2 Effect of Crossover on a Schema

We have seen how selection influences a schema. Now we will talk about how crossover affects various schemata. Assume that there are two schemata in a population at time *t*. Let's call these schemata 100*0**** and 001111****, respectively. Let us examine how a crossover operator influences these schemata.

Let us suppose that 100*0**** is represented by schema $S_i(t)$ and 001111**** S_i is represented by schema $S_{i+1}(t)$.

When a crossover is done on schema 1, the cut point determines whether the same schema will survive. If a cut point is picked after fixed bits, the chances of a schema surviving are quite good. The schema will be destroyed if the cut point is chosen between two fixed numbers of bits. To figure out the chances of surviving, we must first figure out the chances of being destroyed. If each solution is made up of N bits, the cross-site can be evenly chosen from $N - 1$ bits.

$$Probability\,of\,Destructing\,a\,Schema = \frac{\Delta(S_i(T))}{N-1} \tag{2.9}$$

$$The\,Probability\,of\,Surviving\,A\,Schema = 1 - \frac{\Delta(S_i(T))}{N-1} \tag{2.10}$$

If the probability of crossover is P_c, then the chances of survival will be

$$\frac{Pc * \delta(S_i(t))}{n-1} \tag{2.11}$$

Even though we anticipated there would be no possibility of survival if the cut point were in the fix bits, this assumption isn't correct. Even after selecting a cut point between fixed bits, the schema may survive in some circumstances. Considering all these factors, the likelihood of a schema surviving selection and crossover can be defined by Equation 2.12.

$$likelihood\ of\ surviving\,a\ schema \geq S_i(T) * F(T) * \frac{*\left(1 - \mathrm{Pc} * \Delta(S_i(\mathrm{T}))/\mathrm{N}-1\right)}{Average\,Fitness\,of\,Population} \tag{2.12}$$

where, $F(T)$ = Fitness of Schema at Time T.

2.4.3 Effect of Mutation on the Survival of a Schema

Crossover and mutation both produce new offspring. Again, let us take one example and see the chances of survival for a schema after mutation is implemented.

Let us take the same schema 100*0*****. After mutation, the changes in the schema will take place only if fixed bits get flipped if the probability of mutating one bit is P_m, then the probability of survival is $(1 - P_m)$ and the probability of surviving a schema after mutation will be

$$(1-P_m)o(S_i(t)) \geq 1 - o(S_i(t)) * P_m. \tag{2.13}$$

So finally, the chances of survival of a schema after selection, crossover, and mutation will be

$$S_i(t) * population\,size * fitness\,of\,schema\,at\,time\,t * \left(1 - P_c * \frac{\delta(S_i(t))}{n-1}\right) * \frac{\left(1 - o(S_i(t)) * P_m\right)}{average\,fitness\,of\,population}.$$

This theoretical discussion shows that the chances of survival of low-order schema will be very high.

So ultimately, a GA promotes a schema with a low order.

2.5 MATLAB® PROGRAM FOR A GA

The MATLAB programming language offers extensive support for solving optimization problems using the GA. It includes a tool for solving optimization problems with a GA. For producing an optimal solution using a GA by this tool, you must enter some critical information about the problem into this tool. For example, the tool will request that you provide a fitness function. It will also inquire about the type of selection, crossover, and mutation operators used. We have talked about a variety of GA operators. You are free to use any of these operators to solve your problem. When you enter all required information, it will automatically generate the best solutions.

People who are not very good at writing programs can benefit from this tool. MATLAB provides an excellent platform for writing codes. Because this book is not about MATLAB programming, it will not explain the syntax and semantics of each instruction defined in MATLAB programming. The syntax of MATLAB programming is very similar to that of C programming. If you know how to write a C program, you can quickly write a MATLAB program. Like the C programming language, MATLAB provides all necessary support for defining control statements and functions.

We are assuming that the reader has some basic knowledge of programming language. They can design some functions and know about looping and other constructs. In upcoming section, we discuss how the GA program can be written and implemented.

Writing a good program is an art. A good program can be easily maintained and extended. Procedural programming and object-oriented programming are two popular approaches to writing good programs. MATLAB supports both types of programming. We will use the procedural method to write the GA program.

2.5.1 Writing Program for a Binary GA

In procedural programming, everything in the program is written in function. A function is written for implementing some functionality by converting some input information into output information. For writing a program of binary GA (BGA), we can create a BGA function. The BGA function will convert an input population into an output population. As we know, conversion of input population into output population is done with the help of selection, crossover, and mutation operators. So ultimately, the function BGA will be created by supplying some input data in the input population. These input data will be converted into output population with the help of operators.

Moreover, as each operator converts an input population to an output population, these operators can be designed as independent functions. So the purpose of the BGA function can be shown using the following function.

Output population Function BGA (input parameters)

```
{
Input population;
Function SelectionOperator(input parameter);
Function CrossoverOperator(input parameter);
Function MutationOperator(input parameter);
}
```

So for writing the code of BGA, the declaration of the BGA function is the first step in writing the BGA program. The BGA function will take some of the input population and generate the output population. All input parameters defined in the GA program will be either passed from another function or declared within the same function. Population size, crossover probability, mutation probability, and other parameters needed to encode solutions are essential parameters in a GA.

In MATLAB, the declaration of a variable is straightforward. Any name can be used to declare any variable. For example, for expressing population size, we will use `psize`. You can also rename it to `ps`. The name of the variable is dependent on your ease of understanding. You can change the name of any variable according to your requirement. You do not need to mention the type of the variable. Any variable can be declared by assigning some value to it. After defining every variable, you have to terminate it with the semicolon.

The code for defining these important variables is listed in Table 2.17

TABLE 2.17
Defining Important Input Variables for Function BGA

`maxit=2;`	for defining the max number of iterations, how many times do you repeat GA
`fprintf ("Min cost");`	Used to display the strings written under fprintf, used for better readability, before displaying minimum cost print string Min Cost
`mc=-100 or any other suitable value;`	initial minimum cost
`psize=10 ;`	population size (we can also assign another name to population size like ps)
`mut=.15 ;`	mutation rate
`sel=0.5;`	fraction of population passing through crossover and mutation
`fprintf("number of bits per parameter") ;`	displays string number of bits per parameter
`str=8;`	number of bits in each parameter
`fprintf("total number of bits");`	display string total number of bits
`dim = 3;`	dimension of solution
`nbits = str*dim ;`	total number of bits required to represent the whole solution
`fprintf("POP member that survive");`	surviving population
`ns = floor(selection*psize)`	how many solutions will participate in crossover and mutation

When you run GA code, the following values will be printed:

Max iterations are declared by the variable

```
maxit = 2
```

Similarly, other values of variables will be initialized like Min cost

```
mcost = -100
```

psize will define the size of the population, and it will be assigned to

```
psize = 10
```

Mutation probability will be taken as .15.

```
mut = 0.15
```

Similarly, the selection probability will be .5.

```
sel = 0.5
```

How can we decide the size of the initial population?
After declaring the BGA function, we must supply the solutions in binary format.

In MATLAB, the data can be transferred directly into a matrix. So we can use some predefined commands for generating data in a single go.

Let us suppose that our input population contains ten solutions, each having 24 bits. How can we generate the whole population by a set of commands? Who will decide that the size of the initial population will be 10? How the number of bits per solution will be determined. We explain everything one by one and show what commands will be required to generate the initial population.

In most of the problems, the programmer determines the size of the initial population. In our situation, it is defined as 10 by declaring the psize variable. Moreover, the bits required for representing the solution will be determined with the help of the dimension of each solution. Dimension represents the number of variables used to create a solution vector. Based on the objective function, we can determine the dimension of the solution of the problem.

Let us take an example.

If the problem is

$$Maximize\, F(x_1,x_2,x_3) = (x_1^2 + x_2^2 + x_3^2),$$

$$where\, 1 \le x_1 \le 5, 2 \le x_2 \le 6, 1 \le x_3 \le 7,$$

then each solution will be composed of three variables `x1, x2` and `x3`. So the dimension of the solution of this problem will be 3.

To represent this solution in binary, we have to define how many bits we have to use to represent each solution variable. These bits are defined as the number of bits per parameter.

The number of bits per parameter: These bits represent the number of bits required to represent each variable. In the preceding example, we have used 8 bits to represent one input variable. Therefore, `str` will define the number of bits required to represent each variable in the input.

```
str = 8;
```

In the example problem, each variable `x1, x2` and `x3` is represented by `8` bits.

`nbits` variable will calculate the total number of bits in a solution, and for a three-variable solution, the total number of bits will be

```
nbits = 24
```

dim is 3 as each solution is made up of 3 variables.

psize will define the size of the population, and it will be assigned to

```
psize = 10
```

As we have already discussed in the theory of GA, the first step of a GA is to initialize the population. Now we use some code for generating the initial population.

2.5.1.1 Generating the Initial Population

An initial population of the BGA for a given optimization problem is represented by a set of initial binary solutions for the problem. The programmer defines the number of binary solutions to represent the initial population. Before defining the solution in binary, we must first determine how many bits will be used to represent each solution. The number of binary bits taken to represent each solution is dependent on the dimension of the problem. In our example, each solution comprises three input variables, so each solution is represented by a combination of three input variables. The total number of bits required to represent each solution can be calculated by multiplying the number of bits taken to represent each variable and the number of variables used to represent the solution. In most problems, these input variables are characterized by a dimension variable. For example, the solution to a two-variable problem will be represented by a combination of two input variables, and the dimension of the solution in this problem will be 2. By multiplying the dimension by the number of bits per dimension, the total number of bits needed to represent the solution is calculated.

Let us look at an example of how to generate a binary population in MATLAB.

Before displaying the initial population, we can use `fprintf("Initial Population");` to print what we will display currently. The `fprintf` command in MATLAB allows you to display the string written inside the `fprintf` command.

So `fprintf("Initial Population");` will display string "Initial Population"

After displaying, initial population, we have to print the initial population.

As we have seen, the initial population size is 10, and each solution has 3 variables. There is a need to produce 10 solutions of each 24 bits.

To produce the whole population in one go, we first generate 10 rows each containing 24 random values with the help of `rand(psize, nbits)` command.

Would you mind noticing that psize is 10 and nbits is 24? Each value generated by this command will be an arbitrary value between 0 and 1. Finally, to convert these values into binary, we have to apply a `round` command. This `round` will convert a value to 1 if greater than .5 else to 0. So command `pop=round(rand(psize,nbits)` will generate binary bits for the input Population.

MATLAB allows users to generate a whole population by using simple commands easily. It also provides power to the users to directly work with matrices. For example, in MATLAB, only a single instruction can generate ten solutions with 24 bits. This can be done by using the command `round(rand((psize, nbits))`. The `rand(psize, nbits)` command generates 240 random numbers, each less than 1, in 10 rows. Each row of this matrix represents 24 random numbers. These will be fractional values, with each value ranging from 0 to 1. The round function will be applied to the `rand` function to convert this matrix into binary bits and convert these results into a binary population.

MATLAB code and details for generating initial population:
Command `disp(pop)` will display the population represented by binary strings.

Following is the code for generating the population in binary.

`fprintf("Inital Population");` % will display string initial population

`pop=round(rand(psize,nbits));` % this will be used to generate binary bits for whole population.

`disp(pop);` % display population

These three statements will define the initial population and display it as binary numbers. Each solution is represented by 24 bits, and there will be ten of them.

An initial sample population consisting of 10 rows with 24 bits is shown in Table 2.18.

Converting binary values into real values:

Following the generation of binary values, these values will be converted to real values. This coverage is performed to compute the fitness value of each solution. The fitness calculation will be used to apply the GA selection operator.

The following code will convert a binary string to its corresponding real value. Assume that the lower bound of each variable is a, and the upper bound is b. Then, each variable can be represented in real value using the code that follows.

Before representing the solution in real numbers, we must first determine its dimension. How many bits are used to describe one variable? Because each variable

TABLE 2.18
Randomly Generated Binary Solutions

0	0	0	0	0	0	1	1	0	0	0	1	0	0	1	0	1	1	1	0	0	1	0	1
1	0	1	0	0	1	1	0	1	0	0	0	0	0	1	1	0	0	0	0	0	1	0	1
1	0	1	1	1	0	1	1	1	1	0	0	0	0	1	1	0	0	0	1	1	1	1	0
0	1	0	0	0	0	1	1	1	0	1	0	1	0	0	1	0	0	0	0	0	1	1	1
1	1	0	1	1	0	1	1	1	0	1	1	0	1	0	1	0	0	0	1	0	1	0	1
1	1	1	0	0	0	0	1	1	1	0	0	0	0	1	1	1	1	0	1	1	1	0	0
0	1	0	1	1	1	0	0	0	1	0	0	0	1	0	1	1	1	0	0	1	1	1	1
1	1	1	0	1	0	0	0	1	1	1	0	0	0	1	1	1	0	0	1	0	0	1	1
0	0	0	0	1	1	1	0	1	0	0	1	1	0	0	1	0	1	0	0	0	0	0	0
1	0	0	0	1	1	1	1	0	1	1	1	0	1	1	0	1	1	1	1	1	1	0	1

is represented by `str` bits. We divide the total number of bits by the size of the solution. In this way, we can determine which binary string will be used for which variable. Then, using a command, we convert these variables from binary to decimal and then from decimal to real values.

Command '`bi2de`' is used to convert binary numbers to decimal numbers.

The real value of *i*th variable can be calculated by command.

act variable is taken to represent actual value of the solution.

```
act(:, i)=((b-a)/(2^str-1)*(dec(:, i))+a);
```

The following program can display the process of conversion from binary to decimal and displays it:

```
for i=1:dim
dec(:,i)=bi2de(pop(:,(i-1)*nbits+1:i*nbits));
act(:,i)=((b-a)/(2^strlen-1)*(Dec(:,i))+a);
end
```

Let us take an example and see how it converts a binary value to actual value. Take the first solution from the binary population.

0	0	0	0	0	0	1	1	0	0	0	1	0	0	1	0	1	1	1	0	0	1	0	1

This is the binary vector since we know that three variables represent each solution.

The first 8 bits will be used for the first variable, the second 8 bits will be used for the second variable, and third 8 bits will be used for third variable.

First, these three binary strings will be converted into decimal values by command `dec(:,i)=bi2de(pop(:,(i-1)*nbits+1:i*nbits));`

So decimal value for first variable will be 3.

So `dec(:,1)`=decimal value of `(00000011)=3`

Similary `dec(:,2)`=decimal value of `(00010010)`

And `dec(:,3)`=decimal value of `(11100101)`

Furthermore, these decimal values will be converted into actual values by taking lower and upper bounds.

Binary to real conversions

```
0.11765    0.70588    8.9804
6.5098     5.1373     0.19608
7.3333     7.6471     1.1765
2.6275     6.6275     0.27451
8.5882     7.098      0.82353
8.8235     7.6471     8.6275
3.6078     2.7059     8.1176
9.098      8.902      5.7647
0.54902    6          2.5098
5.6078     4.6275     9.9216
```

Now we have created the initial population. To apply selection, crossover, and mutation, we have to check the fitness of solutions.

The fitness of each solution can be evaluated by using `feval` command. `ff` will be the fitness function.

This code will be used to print the fitness

```
for i=1:10
cost(i)=feval(ff,act(i)); % calculates population
end
```

Calculate fitness

```
Columns 1 through 10
```

0.013841	42.378	53.778	6.9035	73.758	77.855	13.017	82.774	0.30142	31.448

After determining the fitness of solutions, they can be arranged to meet the needs of operators. In general, the selection operator assigns solutions based on their fitness. Based on each solution's fitness value, we can place them in ascending order. As a result, the solution with the least fitness will be stored first, and the solution with the most fitness will be stored last. In addition to cost, their position should be arranged based on fitness.

The following command can be used to accomplish this.

`[cost,ind]=sort(cost)` % this command will arrange all solutions according to their fitness in ascending order with min cost in element 1

```
cost =
```

0.013841	0.30142	6.9035	13.017	31.448	42.378	53.778	73.758	77.855	82.774

```
ind =
```

1 9 4 7 10 2 3 5 6 8

This command will place the minimum value at the top of the list, and the other values will be arranged accordingly, with their indices placed in the index value.

After all solutions have been arranged, their positions can be reordered using new indices.

`act=act(ind,1:3)` % positions represented as actual values are displayed by act

`pop=pop(ind,1:24)` % sorts binary population with % lowest cost first

New positions ordered according to fitness

```
act =

    0.11765     0.70588     8.9804
    0.54902     6             2.5098
    2.6275      6.6275      0.27451
    3.6078      2.7059      8.1176
    5.6078      4.6275      9.9216
    6.5098      5.1373      0.19608
```

TABLE 2.19
Sorted Binary Population according to Fitness

0	0	0	0	0	0	1	1	0	0	0	1	0	0	1	0	1	1	1	0	0	1	0	1
0	0	0	0	1	1	1	0	1	0	0	1	1	0	0	1	0	1	0	0	0	0	0	0
0	1	0	0	0	0	1	1	1	0	1	0	1	0	0	1	0	0	0	0	0	1	1	1
0	1	0	1	1	1	0	0	0	1	0	0	0	1	0	1	1	1	0	0	1	1	1	1
1	0	0	0	1	1	1	1	0	1	1	1	0	1	1	0	1	1	1	1	1	1	0	1
1	0	1	0	0	1	1	0	1	0	0	0	0	0	1	1	0	0	0	0	0	1	0	1
1	0	1	1	1	0	1	1	1	1	0	0	0	0	1	1	0	0	0	1	1	1	1	0
1	1	0	1	1	0	1	1	1	0	1	1	0	1	0	1	0	0	0	1	0	1	0	1
1	1	1	0	0	0	0	1	1	1	0	0	0	0	1	1	1	1	0	1	1	1	0	0
1	1	1	0	1	0	0	0	1	1	1	0	0	0	1	1	1	0	0	1	0	0	1	1

```
    7.3333       7.6471       1.1765
    8.5882       7.098        0.82353
    8.8235       7.6471       8.6275
    9.098        8.902        5.7647

The same positions in binary are represented by pop

Pop= Columns 1 through 24

Sort the Population using the cost

    0.013841       1
    0.30142        9
    6.9035         4
    13.017         7
    31.448         10
    42.378         2
    53.778         3
    73.758         5
    77.855         6
    82.774         8
```

Now we will look at how to use the selection operator. However, several selection methods can be used to select solutions. We use the roulette wheel selection method in this program.

To generate a new intermediate population, we first determine how many solutions will participate in the selection.

2.5.1.2 Implementing Selection Operator in a BGA Using MATLAB

After generating the initial population, the programmer will decide how many solutions will produce offspring. In some cases, only a few solutions in the population will produce offspring. Few of the best solutions will not produce offspring and will be passed down from one generation to another. Such GA programs are referred to as elitist GAs. In elitist GA, Good solutions from previous generations are kept in separate storage for future generations. They will not participate in producing offspring as if they participate in producing offspring.

```
ns=floor(selection*psize)
M=ceil((popsize-ns)/2);
```

The value of the M variable will determine the number of matings performed by a pair of solutions. The number of nonparticipating solutions (ns) is 5. M will be used to define the number of mating. In this program, the number of matings is taken as 3.

Mating will be implemented in pairs, and for every mating, two solutions will be selected randomly from the population since roulette wheel selection required calculations of cumulative probability.

The following code will be used to define the selection.

For implementing roulette wheel selection, the portion of the fitness wheel covered by each solution can be determined by the following command

`prob=flipud([1:ns]'/sum([1:ns]));` % this will determine the proportion of each solution in defining the roulette wheel; after checking the weightage of each solution, the wheel can be arranged according to the fitness of solutions.

```
sum([1:ns]=15
transpose of [5/15 4/15 3/15 2/15 1/15]
```

Probability distribution function:
Odds are calculated by adding a new value with the previous value. Starting with 0, the next value will be 0.3333. Similarly, the next value will be .3333+.2667=.6000. Similarly, other values can be calculated.

TABLE 3.7
Chromosome Weights

0.3333
0.2667
0.2000
0.1333
0.0667

TABLE 3.8
Table Showing Cumulative Odds

0
0.3333
0.6000
0.8000
0.9333
1.0000

```
odds=[0 cumsum(prob(1:ns))']; % probability distribution function
```

Once these values are created, the solution that participated in each mating can be determined by the following code.

For the identification of the solution that will participate in mating, three random numbers are generated for the first solution, and three random numbers are generated for the second solution.

To define the solution index, we check which solutions will cover these random numbers. In each mating, those solutions will be selected for mating.

```
fix1=rand(1,M); % mate #1
fix 2=rand(1, M); % mate #2
```

Now for defining the index of solution, we check which solutions will be covering these random numbers. Code for selecting solutions for mating follow:

```
fix1=rand(1,M); % Mate 1
fix2=rand(1,M); % Mate 2
```

Mate 1

```
fix1 = rand(1,M) = {0.5999, 0.5643, 0.9900}
```

Mate 2

```
fix2 = rand(1,M) = {0.7466, 0.4118, 0.9289}
count=1;
while count<=M
for id=2:ns+1
if fix1(count)<=odds(id) & fix1(count)>odds(id-1)
ma(count)=id-1;
end % if
if fix2(count)<=odds(id) & fix2(count)>odds(id-1)
pa(count) =id-1;
end % if
end % id
count=count+1;
end
```

This code is used to select a solution for performing crossover. According to the code, the number of matings will be 3. In each mating, two solutions will participate in crossover and mutation operation. Index values of these solutions are identified based on cumulative fitness.

Two solutions selected for mating will be decided randomly based on cumulative probability.

Let us see how this program works.

This code is used to select the solution for performing the crossover. According to the code, the number of mating will be 3. In each mating, two solutions will participate in crossover and mutation operation. Index values of these solutions are identified on the basic cumulative fitness.

Two solutions that are selected for mating will be decided randomly on the basis of cumulative probability.

ma and pa contain the indices of the chromosomes that will mate

```
count = 1
ma = {2}
pa = {3}
count = 2
pa = {3,2}
ma = {2,2}
count = 3
ma = {2, 2, 5}
pa = {3, 2, 4}
```

Chromosome selected for mating
2 2 5
3 2 4

2.5.1.3 Implementing One-Point Crossover in MATLAB

After selection, the one-point crossover is produced to generate offspring solutions. A cut will determine the nature of offspring. If the cut point is near the least significant bit, exploitation can be performed. Similarly, the crossover operator will explore the search space if the cut point is close to the most significant bit. After crossover, the selected population will be converted into the following population.

The size of the solutions will participate in crossover operation.

Single-point crossover of population: MATLAB code for implementing the single-point crossover follows.

In every mating, the cut point will be decided randomly, and then two new offspring will be created. The same operation will be implemented in all mating.

```
ix=1:2:ns;

crop=ceil(rand(1, M)*(tb-1)); % crossover point

pop(ns+ix, :)=[pop(ma,1:crop) pop(pa,crop+1:tb)] % first offspring

pop(ns+ix+1, :)=[pop(pa,1:crop) pop(ma,crop+1:tb)] % second offspring
```

Let us see how first offspring is generated. When rand(1,M) is used, three random values are generated. Let us suppose these values are (0.55, 0.6, 0.7). When first cut point is taken, it will be 0.55 * 23 = 18.

New offspring will replace the sixth and seventh binary strings. As per commands, the sixth offspring will be created by combining 13 bits of the second parent with 11 bits of the third parent.

After crossover, the following population will be generated. So the table is updated after the crossover. The updated solutions are mentioned by bold bits.

```
pop =
Columns 1 through 24
```

TABLE 2.20
Population Generated after Crossover

0	0	0	0	0	0	1	1	0	0	0	1	0	0	1	0	1	1	1	0	0	1	0	1
0	0	0	0	1	1	1	0	1	0	0	1	1	0	0	1	0	1	0	0	0	0	0	0
0	1	0	0	0	0	1	1	1	0	1	0	1	0	0	1	0	0	0	0	0	1	1	1
0	1	0	1	1	1	0	0	0	1	0	0	0	1	0	1	1	1	0	0	1	1	1	1
1	0	0	0	1	1	1	1	0	1	1	1	0	1	1	0	1	1	1	1	1	1	0	1
0	**0**	**0**	**0**	**1**	**1**	**1**	**0**	**1**	**0**	**0**	**1**	**1**	**0**	**0**	**1**	**0**	**0**	**0**	**0**	**0**	**1**	**1**	**1**
0	**0**	**0**	**0**	**1**	**1**	**1**	**0**	**1**	**0**	**0**	**1**	**1**	**0**	**0**	**1**	**0**	**1**	**0**	**0**	**0**	**0**	**0**	**0**
1	0	1	1	1	0	1	1	1	1	0	0	0	0	1	1	0	0	0	1	1	1	1	0
1	1	1	0	0	0	0	1	1	1	0	0	0	0	1	1	1	1	0	1	1	1	0	0
1	0	0	0	1	1	1	1	0	1	1	1	0	1	1	1	1	1	0	0	1	1	1	1

Note: The black and bold bit pattern represents the crossover has been done between the two.

2.5.1.4 Implementing Mutation Operator in MATLAB

After crossover, the mutation operator generates the offspring population. During mutation, first, the bits that are to be flipped will be identified, and then the mutation will be applied. Bits to be mutated are represented in terms of rows and columns. For example, the first mutation is used on 23rd bit of ninth row. Similarly, the second mutation is devoted to 15th bit of second row.

```
fprintf("Bits to be mutated in terms of rows and cols" )
   nmut=ceil((psize-1)*tb*mutrate); % total number of mutations
   mrow=ceil(rand(1,nmut)*(psize-1))+1; % row to mutate
   mcol=ceil(rand(1,nmut)*Nt); % column to mutate
disp(mrow);
disp(mcol);
for ii=1:nmut
pop(mrow(ii),mcol(ii))=abs(pop(mrow(ii),mcol(ii))-1);
Columns 1 through 20
```

9	2	5	9	5	3	9	2	5	6	3	10	4	9	5	5	3	4	9	2

```
Columns 21 through 33
```

8	3	5	9	9	10	7	4	2	7	8	9	10

```
Columns 1 through 20
```

23	15	23	18	11	3	10	4	9	16	10	19	3	23	5	24	13	13	21	2

```
Columns 21 through 33
```

7	17	23	24	15	19	8	10	13	23	5	15	18

In total, 33 mutations are to be implemented. These mutations will be implemented in the population produced after crossover. You can see in total 33 values are generated both for rows and columns. The position of bit that is to be mutated after combining row value to column value. The first value to be mutated will be stored `ar(9,23)`. A new population is generated after mutation, as follows.

```
Columns 1 through 24
```

0	0	0	0	0	0	1	1	0	0	0	1	0	0	1	0	1	1	1	0	0	1	0	1
0	1	0	1	1	1	1	0	1	0	0	1	0	0	1	1	0	1	0	0	0	0	0	0
0	1	1	0	0	0	1	1	1	1	1	0	0	0	0	1	1	0	0	0	0	1	1	1
0	1	1	1	1	1	0	0	0	0	0	0	1	1	0	1	1	1	0	0	1	1	1	1
1	0	0	0	0	1	1	1	1	1	0	1	0	1	1	0	1	1	1	1	1	1	0	0
0	0	0	0	1	1	1	0	1	0	0	1	1	0	0	0	0	0	0	0	0	1	1	1
0	1	0	0	0	0	1	0	1	0	1	0	1	0	0	1	0	1	0	0	0	0	1	0
0	0	0	0	0	1	0	0	1	0	0	1	1	0	0	1	0	1	0	0	0	0	0	0
0	0	0	0	1	1	1	0	1	1	0	1	1	0	0	1	0	0	0	0	1	0	0	1
1	0	0	0	1	1	1	1	0	1	1	1	0	1	1	1	1	0	0	0	1	1	1	1
0	1	0	1	1	1	0	0	0	1	0	0	0	1	0	0	1	1	1	1	1	1	0	1

After mutation new population is generated, we reevaluate this population and check the fitness of new solutions.

Since the offspring population is binary, it should be converted into real values before reevaluation.

After selection, crossover, and mutation, new offspring are generated. To check whether we have reached to the desired optimal solution. The fitness of the whole population is reevaluated.

```
par =
        0.11765      0.70588       8.9804
        3.6863       5.7647        2.5098
        3.8824       8.8235        5.2941
        4.8627       0.5098        8.1176
        5.2941       8.3922        9.8824
        0.54902      5.9608        0.27451
        2.5882       6.6275        2.5882
        0.15686      6             2.5098
        0.54902      8.5098        0.35294
        5.6078       4.6667        5.6078
```

```
After generating the new population, another cycle of crosso-
ver and mutation operator is applied.

pop =

Columns 1 through 24
```

0	0	0	0	0	0	1	1	0	0	0	1	0	0	1	0	1	1	1	0	0	1	0	1
0	1	0	1	1	1	1	0	1	0	0	1	0	0	1	1	0	1	0	0	0	0	0	0
0	1	1	0	0	0	1	1	1	1	1	0	0	0	0	1	1	0	0	0	0	1	1	1
0	1	1	1	1	1	0	0	0	0	0	0	1	1	0	1	1	1	0	0	1	1	1	1
1	0	0	0	0	1	1	1	1	1	0	1	0	1	1	0	1	1	1	1	1	1	0	0
0	0	0	0	1	1	1	0	1	0	0	1	1	0	0	0	0	0	0	0	0	1	1	1
0	1	0	0	0	0	1	0	1	0	1	0	1	0	0	1	0	1	0	0	0	0	1	0
0	0	0	0	0	1	0	0	1	0	0	1	1	0	0	1	0	1	0	0	0	0	0	0
0	0	0	0	1	1	1	0	1	1	0	1	1	0	0	1	0	0	0	0	1	0	0	1
1	0	0	0	1	1	1	1	0	1	1	1	0	1	1	1	1	0	0	0	1	1	1	1

```
Next second crossover with mutation is applied.
ans =
       1      0.013841
Chromosome selected for mating
1  2  4
1  1  1
Single point crossover of population
pop =
Columns 1 through 24
```

0	0	0	0	0	0	1	1	0	0	0	1	0	0	1	0	1	1	1	0	0	1	0	1
0	1	0	1	1	1	1	0	1	0	0	1	0	0	1	1	0	1	0	0	0	0	0	0
0	1	1	0	0	0	1	1	1	1	1	0	0	0	0	1	1	0	0	0	0	1	1	1
0	1	1	1	1	1	0	0	0	0	0	0	1	1	0	1	1	1	0	0	1	1	1	1
1	0	0	0	0	1	1	1	1	1	0	1	0	1	1	0	1	1	1	1	1	1	0	0
0	0	0	0	0	0	1	1	0	0	0	1	0	0	1	0	1	1	1	0	0	1	0	1
0	1	0	0	0	0	1	0	1	0	1	0	1	0	0	1	0	1	0	0	0	0	1	0
0	1	0	0	0	0	1	1	0	0	0	1	0	0	1	0	1	1	1	0	0	1	0	1
0	0	0	0	1	1	1	0	1	1	0	1	1	0	0	1	0	0	0	0	1	0	0	1
0	1	1	0	0	0	1	1	0	0	0	1	0	0	1	0	1	1	1	0	0	1	0	1

```
pop =
Columns 1 through 24
```

0	0	0	0	0	0	1	1	0	0	0	1	0	0	1	0	1	1	1	0	0	1	0	1
0	1	0	1	1	1	1	0	1	0	0	1	0	0	1	1	0	1	0	0	0	0	0	0
0	1	1	0	0	0	1	1	1	1	1	0	0	0	0	1	1	0	0	0	0	1	1	1
0	1	1	1	1	1	0	0	0	0	0	0	1	1	0	1	1	1	0	0	1	1	1	1
1	0	0	0	0	1	1	1	1	1	0	1	0	1	1	0	1	1	1	1	1	1	0	0
0	0	0	0	0	0	1	1	0	0	0	1	0	0	1	0	1	1	1	0	0	1	0	1
0	0	0	0	0	0	1	1	0	0	0	1	0	0	1	0	1	1	1	0	0	1	0	1
0	1	0	0	0	0	1	1	0	0	0	1	0	0	1	0	1	1	1	0	0	1	0	1
0	0	0	1	1	1	1	0	1	0	0	1	0	0	1	1	0	1	0	0	0	0	0	0
0	1	1	0	0	0	1	1	0	0	0	1	0	0	1	0	1	1	1	0	0	1	0	1
0	0	0	1	1	1	0	0	0	0	0	0	1	1	0	1	1	1	0	0	1	1	1	1

```
Bits to be mutated in terms of rows and cols
Columns 1 through 20
```

5	3	5	6	6	4	3	10	6	8	6	9	7	7	10	3	6	2	5	4

```
Columns 21 through 33
```

7	5	8	4	2	9	10	9	10	10	4	9	7

```
Columns 1 through 20
```

12	22	20	20	2	16	3	5	7	14	6	20	2	22	18	5	17	13	4	11

```
Columns 21 through 33
```

23	5	5	14	7	19	6	4	19	1	22	3	1

```
Population after mutation
Columns 1 through 24
```

0	0	0	0	0	0	1	1	0	0	0	1	0	0	1	0	1	1	1	0	0	1	0	1
0	1	0	1	1	1	0	0	1	0	0	1	1	0	1	1	0	1	0	0	0	0	0	0
0	1	0	0	1	0	1	1	1	1	1	0	0	0	0	1	1	0	0	0	0	0	1	1
0	1	1	1	1	1	0	0	0	0	1	0	1	0	0	0	1	1	0	0	1	0	1	1
1	0	0	1	1	1	1	1	1	1	0	0	0	1	1	0	1	1	1	0	1	1	0	0
0	1	0	0	0	1	0	1	0	0	0	1	0	0	1	0	0	1	1	1	0	1	0	1
1	1	0	0	0	0	1	1	0	0	0	1	0	0	1	0	1	1	1	0	0	0	1	1
0	1	0	0	1	0	1	1	0	0	0	1	0	1	1	0	1	1	1	0	0	1	0	1
0	0	1	0	1	1	1	0	1	0	0	1	0	0	1	1	0	1	1	1	0	0	0	0
1	1	1	0	1	1	1	1	0	0	0	1	0	0	1	0	1	0	0	0	0	1	0	1
0	0	0	1	1	1	0	0	0	0	0	0	1	1	0	1	1	1	0	0	1	1	1	1

```
Population reevaluated for cost
par =
     0.11765     0.70588      8.9804
     3.6078      6.0784       2.5098
     2.9412      8.8235       5.1373
     4.8627      1.5686       7.9608
     6.2353      7.7647       9.2549
     2.7059      0.70588      4.5882
     7.6471      0.70588      8.902
     2.9412      0.86275      8.9804
     1.8039      5.7647       4.3922
     9.3725      0.70588      5.2157
pop =
Columns 1 through 24
```

0	0	0	0	0	0	1	1	0	0	0	1	0	0	1	0	1	1	1	0	0	1	0	1
0	1	0	1	1	1	0	0	1	0	0	1	1	0	1	1	0	1	0	0	0	0	0	0
0	1	0	0	1	0	1	1	1	1	1	0	0	0	0	1	1	0	0	0	0	0	1	1
0	1	1	1	1	1	0	0	0	0	1	0	1	0	0	0	1	1	0	0	1	0	1	1
1	0	0	1	1	1	1	1	1	1	0	0	0	1	1	0	1	1	1	0	1	1	0	0
0	1	0	0	0	1	0	1	0	0	0	1	0	0	1	0	0	1	1	1	0	1	0	1
1	1	0	0	0	0	1	1	0	0	0	1	0	0	1	0	1	1	1	0	0	0	1	1
0	1	0	0	1	0	1	1	0	0	0	1	0	1	1	0	1	1	1	0	0	1	0	1
0	0	1	0	1	1	1	0	1	0	0	1	0	0	1	1	0	1	1	1	0	0	0	0
1	1	1	0	1	1	1	1	0	0	0	1	0	0	1	0	1	0	0	0	0	1	0	1

```
ans =
     2     0.013841
```

The whole process is repeated again and again until either the iterations exhausted or optimal value is achieved.

```
"Optimized function is" "F3"
popsize = 10 mutrate = 0.15 par = 3
#generations=2 best cost=0.013841
best solution
0.11765 0.70588 8.9804
```

3 Real-Parameter Genetic Algorithm

3.1 INTRODUCTION

We covered the theory of simple genetic algorithms (SGAs) in the previous chapter. An SGA is also known as a binary GA (BGA). BGA can quickly solve discrete optimization problems in which a binary number can represent the solution. However, there are several other optimization problems in which binary numbers cannot express the input solutions. In such cases, binary encoding cannot be used and becomes ineffective.

To solve these problems, we must select an encoding that can generate each member of the search space solution. In general, the type of encoding used in a particular situation depends on the problem we want to solve. Permutation-based encoding, for example, is more useful in graph-related problems, such as the traveling salesperson problem. Similarly, real-parameter encoding will be very useful in continuous optimization problems[33].

3.1.1 Problems with Binary Encoding

This chapter discusses real-parameter encoding and explains what changes we have to make in the GA to solve the continuous optimization problem. However, before that, let us see why binary encoding will not efficiently handle continuous optimization problems. Why should we choose real-parameter encoding for solving these problems? Since our whole discussion is based on continuous optimization problems. Let us recall the definition of the continuous optimization problem.

In continuous optimization problems, input variables belong to a continuous input domain. In short, the input domain of *i*th variable *x[i]* can be represented by the interval *[L[i],U[i]]* where *L[i]* and *U[i]* are real numbers and represent the lower bound and upper bound of the *i*th variable.

Let us see one example of a continuous optimization problem shown by Equations 3.1 and Equation 3.2. The objective function of the problem is defined by Equation 3.1. It is a three-variable function in which each input variable of the function is bounded by the real-parameter values in Equation 3.2.

Example 1:

$$Maximize\ \ F(x_1, x_2, x_3) = (x_1^2 + x_2^2 + x_3^2) \tag{3.1}$$

$$where\, 1 \le x_1 \le 5, 2 \le x_2 \le 6, 1 \le x_3 \le 7 \tag{3.2}$$

DOI: 10.1201/9781003313649-3

This problem is an example of a continuous optimization problem. It contains three variables, and all belong to a continuous input domain. When the BGA is applied to solve these problems, its performance will not be good due to the following reasons.

3.1.1.1 Poor Coverage of Search Space

To solve any optimization problem by a GA, we need to identify suitable encoding. An appropriate encoding not only covers all possible solutions in the search space but also helps in searching for an optimal solution in less time. Since an infinite solution can be taken in between the given bounds in a continuous optimization problem, real-parameter encoding will work well. Real-parameter encoding can assign any value to a solution vector and represent any solution from the search domain. However, only defining solutions by real-parameter vectors will not work. We need to redesign the operators of a GA so that they can handle real encoded solutions and search for new solutions effectively.

We can also solve continuous search optimization problems with SGA, but we have to make some changes. As a BGA only accepts solutions in binary format, we need to make some changes in a continuous optimization problem. Binary encoding cannot represent an infinite number of solutions of continuous search space. It can work only with a finite number of solutions that can be represented in 2^n. This shows that a binary encoding technique can be used in solving continuous optimization problems by dividing the continuous search space into 2^n number of solutions. In this way, we can convert a continuous search space into a discrete search space. Remember, these solutions must be at the same distance to cover the entire search space. This division will help in representing each solution by n number of bits. Once decided the continuous search space is converted into a discrete search space containing 2^n solutions and the operators of a BGA are applied to identify the optimal solution.

However, operators of BGA will not be able to search effectively in continuous optimization problems as they will be working with a subset of solutions that may not be arranged as expected. Let us see it in more detail. In the previous paragraph, we have seen that we will not apply BGAs without converting continuous search space

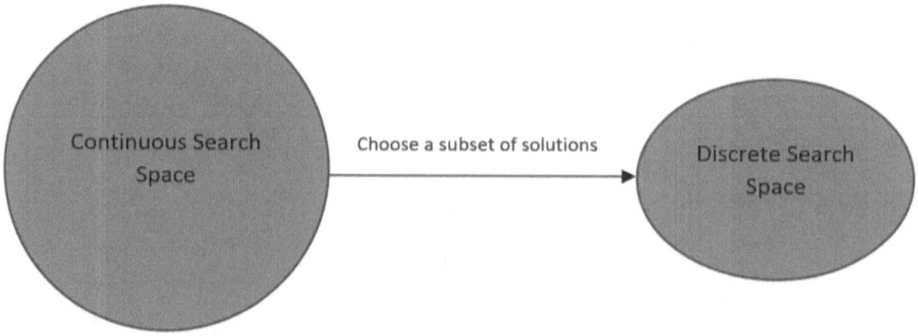

FIGURE 3.1 Converting continuous search space into discrete search space.

into discrete search space. After dividing the search space into a discrete number of points, we can use a BGA. Since we are converting a continuous search space into a discrete number of points, only a proper subset of solutions will be transferred in a discrete search space. Even after choosing a huge value of n, we will not cover the entire solution in a discrete set. If the selected value of n is not sufficiently large, then there will be a very high probability of missing the actual optimum solution in the selected discrete search space. It may be discarded during the conversion from continuous search space to discrete search space. In such a case, even after using a BGA, we will not obtain the exact optimal solution. The search will end up in a local optimal solution. This shows that binary encoding is not good at handling continuous optimization problems. In addition to it, there is one more problem related to the operators. In the BGA, we exploited neighborhood solutions after selecting a good solution, assuming that some other good solutions may lie in the proximity of these solutions. However, when we try these operators with a continuous optimization problem represented by binary encoding, the neighborhood solutions may not be close to the solution (they might be very far from each other). In that case, our search becomes ineffective, and the algorithm will converge lately.

3.1.1.2 More Computational Requirements

A BGA process randomly generates binary solutions to create a new population. The exact process is repeated with the new population until termination criteria are reached. A conversion from an input population to an output population necessitates the selection, crossover, and mutation operators. The selection operator's purpose is to prefer good solutions over bad solutions. The fitness function is used to decide whether a solution is good or bad. So, in each iteration, we must calculate the fitness of each solution.

Because the use of a SGA in continuous optimization problems necessitates using a binary representation of the solution, which is not the case. As a result, we must convert these binary solutions back into actual solutions to test their fitness. This conversion of a binary vector to a real-parameter vector adds to the computation overhead. So, to calculate the solution's fitness value, we must first determine the solution's actual value. This necessitates the conversion of binary values to real-parameter values, imposing a new computational requirement.

3.1.1.3 Longer Computational Time and Lower Accuracy

Because of the increased computational requirements and an inefficient search performed by the BGA operators. Furthermore, a BGA will take more time to achieve an optimal solution because only a subset of solutions is evaluated to identify the optimal solution. A BGA's optimal solution may not be perfect. This demonstrates that BGA requires more computational time and may not produce a very good solution.

3.1.1.4 Hamming Cliff Problem with Binary Encoding

The hamming cliff problem also affects binary encoding, and as a result, the GA operators do not always work as expected. The hamming cliff is associated with the hamming distance (HD) between two solutions. The HD between two solutions can be calculated by comparing the bit changes in two binary strings of equal length.

Example 2: Let us suppose we want to identify the hamming distance between 1100 and 0011.

So HD (1100,0011) = 4 because all bits are changed from 0 to 1 or 1 to 0

Example 3: Take one more example and check the HD between 1100 and 1101

HD (1100,1101) = 1

The HD is beneficial in the design of optimization algorithms. For example, during exploitation, we want to generate new solutions in the vicinity of selected solutions. This can be accomplished by creating an operator that can change one or two bits in the parent vectors. However, due to the hamming cliff problem, a minor change in the parent vector can result in a very distant solution. For example, if the parent vector is 0000 and a one-bit change is made in the most significant bit, the newly generated vector 1000 will be very far from the parent vector.

Similarly, we want to generate new solutions far removed from the parent solution during exploration. So, we try to change many aspects of the parent solution during exploration. Crossover and mutation operators can be used to make these changes. However, even after changing many bits of the parent solutions, the new vector will sometimes be very close to the parent vector. This is also due to the hamming cliff issue. In general, we flip more bits of the parent solution to generate a very different solution from the original one. Still, this method does not always work correctly in some strings. After flipping all the bits, the newly generated solution will be close to the original solution. Due to hamming cliff problem, in several problems, grey encoding is used in place of binary.

3.2 REAL-PARAMETER GA

Due to the problems mentioned earlier, binary encoding will not be efficient in handling continuous optimization problems. In such problems, we have to work with real-parameter encoding. A real-parameter encoding uses a real-valued vector for representing the solution of the problem. However, only defining a solution by a real-parameter vector will not work. To solve continuous optimization problems, we have to redesign the operators of a GA to perform an efficient search. Although the operators' implementation in different versions of GAs is different, the same natural phenomena are implemented in all versions. Also, according to the type of optimization problem, one wants to solve (unimodal or multimodal), the operators of a GA are designed.

In the previous chapter, we explained that a GA works with three operators, that is, selection, crossover, and mutation. The selection operator in a GA creates multiple copies of the good solutions and discards the bad solution. This operator helps in the identification of search areas where the optimal solution may lie. After selection, crossover and mutation operators are used to search for new solutions in the search space. Therefore, we need to modify a real-parameter GA's crossover and mutation operators for efficient searching.

Before discussing the operators of a real-parameter GA in detail, let us look at the working of the GA. A step-by-step explanation follows.

3.2.1 Detailed Implementation of a Real-Parameter GA

Once the encoding is decided, then the operators of a GA can be defined. The ultimate purpose of the GA program is to reach an optimal global solution as early as possible parent population is converted into the offspring population in each iteration by using selection, crossover, and mutation operators. The parent population is generated at random during the first iteration. In subsequent iterations, the offspring population developed in the previous iteration is used as the parent population. Figure 3.2 depicts how the parent population will be converted into the offspring population in each iteration. The flow chart of a GA is shown in Figure 3.2.

Common steps used in all versions of GAs: These are the typical steps common to all GA versions. Let us take some idea of these steps.

Step 1: Initialization of population (parent population)
The initialization of the initial population is the first step in any GA version. The initial population consists of a set of solutions generated at random from the search space. In most cases, initial solutions are chosen at random to cover the entire search space. In general, we choose uniformly distributed solutions, but other distributions, such as Poisson and Gaussian, can also be used depending on the requirements of the problem. The number of solutions in the initial population is set at 40. However, the initial population may include solutions generated by a local strategy for very complex problems or those requiring higher accuracy. This is done to improve the convergence rate of the algorithm. The programmer may also increase the number of solutions.

Step 2: Selection
After the initial population is generated, the selection operator marks good solutions from the population. No new solution is developed during selection. There

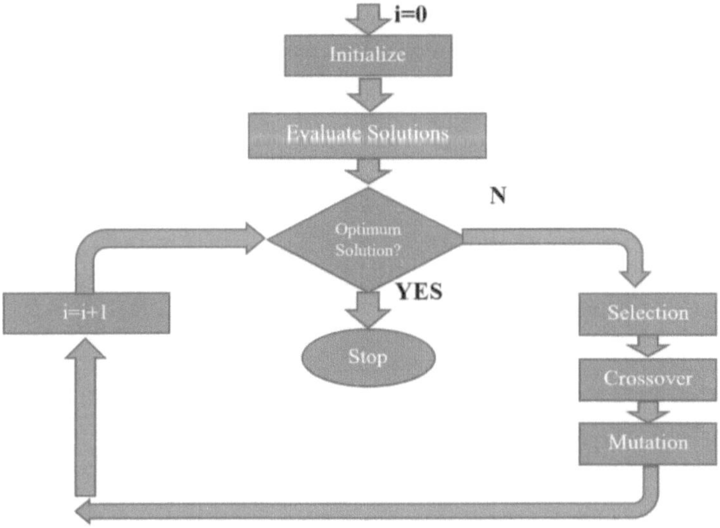

FIGURE 3.2 Flow chart of a GA.

are various methods of selection that are designed to meet the needs of the problem. A selection operator designed to solve unimodal problems, for example, will be ineffective in solving multimodal problems. A selection operator in unimodal problems should make multiple copies of good solutions and discard solutions with the lowest fitness. Only fitness values of solutions should be taken into account when implementing this type of selection. However, this method will not work in a multimodal problem. Different types of selection operators can be designed depending on the type of encoding used to solve the problem or the type of problem to be solved. Always keep in mind that the initial population size will remain constant during selection. Following selection, an intermediate population will be generated.

Step 3: Crossover and mutation (generating the offspring population)
The purpose of the selection operator is to identify good solutions from the initial pool. There will be no new solutions as a result of the selection. However, to solve the problem, we must identify the global optimal solution as soon as possible. Because the initial population comprises randomly generated solutions, there is no guarantee that the global optimal solution will be included in the initial population. To identify the global best solution, we must consider other possible solutions that may be more fit. We must use the exploitation and exploration process to investigate other solutions containing a global optimal solution. During exploitation, new solutions are generated in the vicinity of previously selected good solutions. This is done because the likelihood of retrieving a new good solution close to the already selected solution is very high. However, exploitation alone will not suffice to cover the entire search space. So we must combine the exploration and exploitation processes. Exploration of the whole search space is carried out to identify other good regions for exploitation. So, the entire search space can be covered by combining exploitation and exploration, and the best solution is obtained. The offspring population is created as a result of crossover and mutation. After crossover and mutation, the offspring population is generated. Graphically, the process of the GA can be expressed by Figure 3.3.

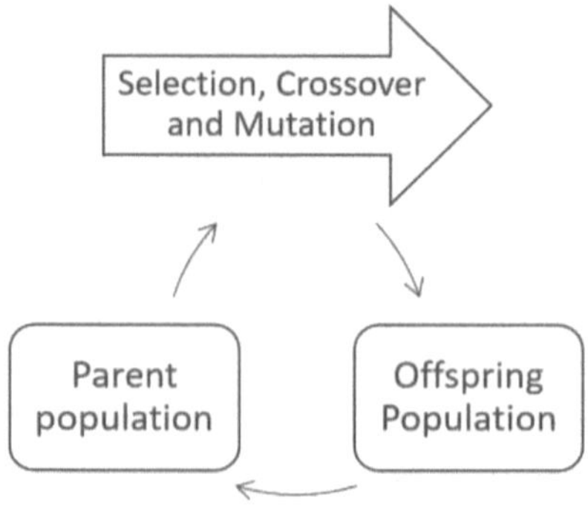

FIGURE 3.3 Working of a GA.

3.2.2 Operators of a Real-Parameter GA

3.2.2.1 Selection Operator

In GA, the selection operator selects a good solution from a population. Because no change is made in the position of the solution during selection, any selection operator described in an SGA can also be implemented in a real-parameter GA. We can use roulette wheel selection, tournament selection, or other methods.

3.2.2.2 Crossover Operator

The purpose of the crossover operator is to generate new solutions in the search space. During the search, we can perform exploitation or exploration. Exploitation is required to search for new solutions in the neighborhood of the already selected solution. Similarly, exploration is necessary to generate a unique solution in the entire search space. A crossover operator in a GA can perform both exploitation and exploration. Let us see some crossover operators used for performing crossover operators in a real-parameter GA.

3.2.2.2.1 Simple Crossover Operator

The most basic crossover operator used in a real-parameter GA is known as a simple crossover operator, which is very similar to the one-point crossover operator used in a BGA. This operator chooses two parents at random from the available solutions to generate offspring. Following that, a cut point is chosen at random from 1 to n – 1, where n is the size of the parent solution. Assume that the real-parameter vectors of parent solutions are represented by X and Y, and their values are 123412 and 678945. A crossover point is chosen at random to perform the crossover operator, and new solutions are generated. Two cut points can also be used in a modified version of the simple crossover operator. Tables 3.1 and 3.2 explain how the crossover operator can be applied. Table 3.1 shows the parent population before crossover, and Table 3.2 shows the offspring population after crossover.

3.2.2.2.2 Linear Crossover Operator

The solution generated by simple crossover may not effectively handle real-parameter solutions. As discussed, it is an extension of the one-point crossover used in

TABLE 3.1
Parents before Crossover

PARENT X	**1234|45**
PARENT Y	**6789|12**

(‘|’ represents the crossover point.)

TABLE 3.2
Offspring after Crossover

CHILD A	**1234|12**
CHILD B	**6789|45**

a BGA. It works effectively with binary values because a combination of 0 and 1 represent each solution. In such a case, when an appropriate cut point is chosen, it will exploit the region correctly. However, in real-parameter encoding, a gene in a chromosome can receive any value from 10 possible digits. In such a case, even after choosing the appropriate cut point, the generated solutions may be very far from the parent vectors. To improve the exploitation ability, several other crossover operators were defined. Linear crossover is the simplest one. A linear crossover operator combines two-parent vector X and Y at position k and generates three vectors. Finally, two vectors with higher fitness are chosen as child vectors. If X is represented by (x_1, . . ., x_n) and Y is represented by (y_1, . . ., y_n), then the operation performed by linear crossover can be summarized by following steps:

Parents: (x1, . . ., xn) and (y1, . . ., yn)

1. Select a random position k.
2. Create three solutions.

   a. $\left(x_1,\ldots,x_k,0.5*y_k+0.5*x_k,\ldots,x_n\right)$

   b. $\left(x_1,\ldots,x_k,1.5*y_k-0.5*x_k,\ldots,x_n\right)$

   c. $\left(x_1,\ldots,x_k,-0.5*y_k+1.5*x_k,\ldots,x_n\right)$

3. The best two are selected based on fitness function and are moved to the next generation from the three children.

3.2.2.2.3 Single Arithmetic Crossover Operator

The single arithmetic crossover operator is a generalized version of the linear crossover operator. The weights of each vector in a single arithmetic crossover vector are determined at random. It should be noted that they were fixed in the linear crossover vector. The operation of a single arithmetic crossover operator is explained in the following steps. The functioning of a single arithmetic crossover operator is defined by the following steps:

1. Parents: (x1, … , xn) and (y1, … , yn)
2. Select a random position k (in this case, it is 4th).
3. Child 1 is created as $\left(x_1,\ldots,x_k,\alpha*y_k+(1-\alpha)x_k,\ldots,x_n\right)$
4. Child 2 will be created by replacing α *with* $(1-\alpha)$. The value of α is 0.5 in this example.

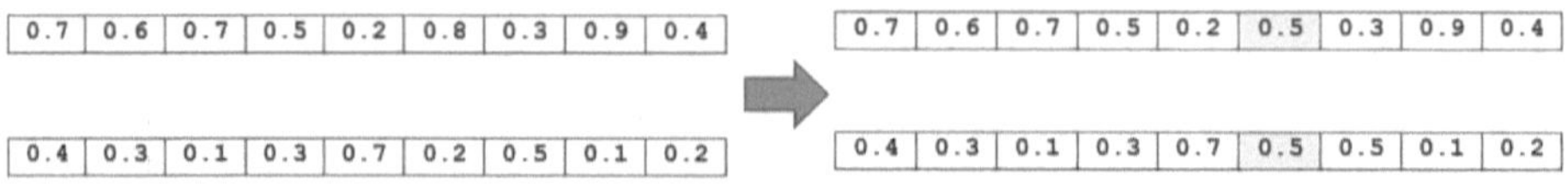

FIGURE 3.4 Example of single arithmetic crossover operator.

3.2.2.2.4 Simple Arithmetic Crossover Operator

The simple arithmetic crossover operator is a generalized version of the single arithmetic crossover operator. Children were generated in single arithmetic crossover by updating the kth position of parent vectors. In the simple arithmetic crossover operator, we change all positions after the chosen random position k. Figure 3.5 explains the simple crossover operator's working.

1. Parents: (x1, … , xn) and (yl, … , yn)
2. Select a random position (k); after this point, mix all values.
3. Child1 is created as $\left(x_1,\ldots,x_k,\alpha * y_{k+1} + (1-\alpha) * x_{k+1},\ldots,\alpha * y_n + (1-\alpha) * x_n\right)$
4. Replace α *with* $(1-\alpha)$ for another child, for example, with $\alpha = 0.5$

3.2.2.2.5 Whole Crossover Operator

This is the most commonly used crossover operator. It is applied to each position of parent vectors. If X and Y are the position vector of parent vectors, then this operator can be implemented as shown in Figure 3.6.

1. *Parents:* $(x_1, \ldots, x_n$) *and* $(y_1, \ldots, y_n)$
2. *Child*$_1$*:* $a.\,\overline{x} + (1-\alpha) * \overline{y}\, a$
3. Child$_2$: $(1-\alpha).\,\overline{x} + (\alpha) * \overline{y}$

3.2.2.2.6 Simulated Binary Crossover Operator

K. Deb and Agrawal propose a simulated binary crossover operator. Unlike another crossover vector, this operator defines some random parameters that can be adjusted to produce the solutions in the targeted regions. If x^1 and x^2 are the two-parent vectors, then a simulated binary crossover can be defined as follows:

$$x_i^{(1,t+1)} = 0.5\left[\left(1+\beta_{qi}\right)x_i^{(1,t)} + \left(1-\beta_{qi}\right)x_i^{(2,t)}\right] \tag{3.3}$$

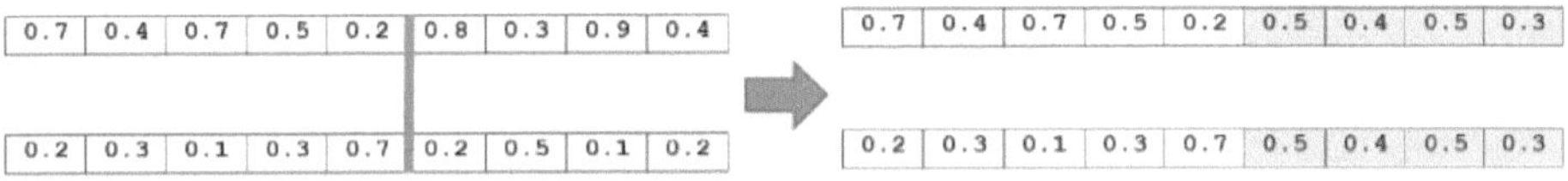

FIGURE 3.5 Example of a simple arithmetic crossover operator.

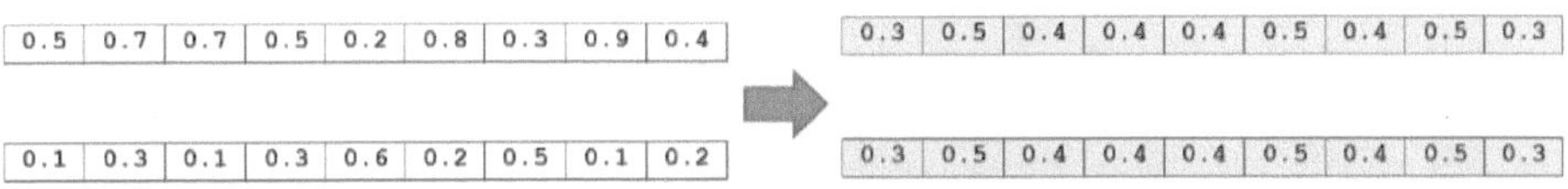

FIGURE 3.6 Example of whole crossover operator.

$$x_i^{(2,t+1)} = 0.5\left[\left(1-\beta_{qi}\right)x_i^{(1,t)} + \left(1+\beta_{qi}\right)x_i^{(2,t)}\right] \tag{3.4}$$

$$\beta_{qi} = \begin{cases} (2u_i)^{\frac{1}{n_c+1}}, & \text{if } u_i \le 0.5, \\ \left(\dfrac{1}{2\left(1-u_i\right)}\right)^{\frac{1}{n_c+1}}, & \text{otherwise} \end{cases} \tag{3.5}$$

where u_i *is* a random number n_c is a parameter that controls the crossover process. A high-value parameter will create a near-parent solution. The overall process used in a simulated binary crossover is shown in Equations 3.3, 3.4, and 3.5.

3.2.2.3 Type of Mutation Operator

In addition to crossover, various mutation operators are available in a real-parameter GA for solving optimization problems. Some most commonly used mutations are described in the following subsections.

3.2.2.3.1 Random Mutation

Random mutation is the simplest method to perturb a solution. It can be used both for performing exploitation and exploration.

$$y_i^{(1,t+1)} = u_i\left(x_i^u - x_i^l\right), \tag{3.6}$$

where u_i is a random number between [0, 1].

$$y_i^{(1,\,t+1)} = x_i^{(1,\,t+1)} + \left(u_i - 0.5\right)\Delta_i, \tag{3.7}$$

where Δ_i is the user-defined maximum perturbation.

Other types of mutations are mentioned in the following subsections.

3.2.2.3.2 Normally Distributed Mutation

A simple and popular method for performing real-parameter mutation is adding normal distribution in the solution. This method is explained in Equation 3.8.

$$y_i^{(1,\,t+1)} = x_i^{(1,\,t+1)} + N\left(0, \sigma_i\right), \tag{3.8}$$

where $N\left(0,\sigma_i\right)$ is the Gaussian probability distribution with a zero mean.

3.2.2.3.3 Polynomial Mutation

In addition to random and normal mutation is polynomial mutation.

Deb and Goyal in 1996 proposed this mutation.

$$y_i^{1,t+1} = x_i^{1,t+1} + \left(x_i^u - x_i^l\right)\delta_i \tag{3.9}$$

$$\delta_i = \begin{cases} (2u_i)^{1/(\eta_m+1)} - 1, & if\, u_i < 0.5 \\ 1-\left(2(1-u_i)\right)^{1/(\eta_m+1)}, & if\, u_i \geq 0.5 \end{cases} \tag{3.10}$$

3.3 MATLAB® PROGRAM FOR A REAL-PARAMETER GA

The MATLAB programming for a real-parameter GA can be written either as a script or a function. MATLAB scripts contain a collection of MATLAB commands and processes used to solve a specific problem. In functional form, the entire program is written in procedures. Where each method transforms input data into output data. Each procedure in MATLAB is defined in the form of a function and has a unique name associated with it, and it takes some input parameters and outputs some output parameters. Whatever method we use to write a real-parameter GA program, the program always begins with declaring important information. This information is required to start and run the GA program. For example, some variables are declared in the GA to start the program and assess solutions. We can initialize this information in one place. Following commands can be used to start the GA.

```
    ff ='testfunction';                        % objective function
    Dim=3;                                     % number of optimization variables
    varhi=xBound; varlo=-xBound;               % variable limits
xBound=5;
```

The `ff` notation is used to initialize the fitness function. This function is required for solution evaluation. A variety of functions can be used to define the `ff` function. A sphere is used to represent a function. It is a scalable function that can be applied to varying size variables. This function is described in documents used to define CEC benchmarks.

At the time of initiation, solution-specific information is required in addition to the fitness function. Real values represent the answer in GA with real parameters. Before building any random solution, we must offer a solution dimension. The number of variables in a solution is referred to as its dimensionality. For each variable, variable bounds should be included. The variable boundaries for all dimensions in our example are the same. `varlo` denotes the variable's lower bound, while `varhi` denotes the variable's upper bound.

These variables can be initialized by giving them a name and assigning them a value. All statements in MATLAB are terminated with a semicolon.

After defining this information, we can specify how many times the GA algorithm must be run, and the cost of the solution must be set to some value.

```
maxit=1000; % max number of iterations
mincost=-9999999; % minimum cost
```

`maxit` defines the maximum number of iterations, and min-cost defines the minimum cost. Alike other variables these variables can also be initialized by defining the name and assigning these variables to some value.

In next declaration, we define parameters related to operator of the GA. Some of these parameters are population size, crossover probability, mutation probability, and selection probability. Let us see the whole set of parameters:

```
popsize=10;                          % set population size
mut=.2;                              % set mutation rate
selection=0.5;                       % fraction of population kept
ns=floor(selection*popsize);         % #population
nmut=ceil((popsize-1) * Dim*mut);    % total number of mutations
M=ceil((popsize-ns)/2);              % members that survive
```

The details of some important parameters and their initialization method are also mentioned in Table 3.3 for ready reference. Please go through it. Each constant or variable vector can be defined by simply writing the name of the variable or constant and assigning some value to it. In MATLAB, each statement should end with a ";" symbol. These parameters are very similar to parameters used in binary GAs.

When you run the GA code, the following values will be printed:

Number of optimization variables
```
Dim = 3
```

Variable limits
```
varhi = 5
varlo = -5
```

Max number of iterations
```
maxit = 1000
```

Population size
```
popsize = 10
```

Mutation rate
```
mut = .2
```

TABLE 3.3
List of the Terms Used in a Real-Parameter GA (RPGA) and Their Descriptions

Terms/Notations	Description
`maxit=1000;`	Variable name for defining max number of iterations, how many times you want to repeat RPGA
`fprintf("Min cost");`	Before displaying minimum cost print string Min Cost
`mincost=--9999999;`	A default variable for defining the initial minimum cost
`popsize=10 ;`	For defining the population size
`mut=.15;`	For defining the mutation rate
`selection=0.5;`	Fraction of population passing through crossover and mutation
`Dim =3;`	Number of input variables in the solution. In this program, it is 3
`fprintf ("POP member that survive");`	Surviving population
`ns=floor(selection*popsize);`	Number of solutions participating in crossover and mutation

Fraction of population kept
`selection = 0.5000`

Population members that survive
`ns = 5`

Total number of mutations
`nmut = 6`

Number of mating
`M = 3`

These parameters are used in each iteration. With one example, we will show how we implement different operators in MATLAB. What is the meaning of each command? We will show one iteration. The same process can be used to generate solutions in other iterations.

3.3.1 Generating the Initial Population

The real-parameter GA should represent the initial population by real-parameter solutions. We have seen that we will be using ten solutions in each generation, each having three dimensions; therefore, we have to check the upper and lower bound for each dimension. After reviewing the bounds, we can use predefined functions of MATLAB for directly generating the whole population. Let us suppose that if each variable is bounded within the interval (–5, 5), this method can generate the initial population.

The steps for initializing initial populations are as follows.

Step 1: generate an initial population of solutions having random values between `0` to `1` by function `rand (r, c)` to create a matrix having `r` rows and `c` columns. Each element of the matrix will be initialized by a random number which will be in between `0` to `1` as shown in Table 3.4.

TABLE 3.4
Random Values Generated between 0 to 1

{0.5283,	0.603,	0.7323}
{0.1002,	0.0716,	0.7933}
{0.4460,	0.9560,	0.2017}
{0.8846,	0.1225,	0.7419}
{0.9448,	0.9042,	0.0877}
{0.6987,	0.6292,	0.9372}
{0.8724,	0.6245,	0.2115}
{0.9560,	0.1362,	0.2286}
{0.2573,	0.3638,	0.7237}
{0.1080,	0.9200,	0.8333}

Create the initial population

```
t1 = rand (popsize, Dim);
```

The random values generated by the rand function can be updated to lie within given bounds. The following command will produce each value within the given bounds:

```
par = (varhi - varlo)* t1 + varlo;
```

After initializing the population, the fitness of each solution can be calculated by the following statements.

Step 2: Evaluate the fitness of each individual
This code will be used to print the fitness:

```
for i=1:10
cost(i)=feval(ff,par(i));
end
```

The initial population, along with fitness values, is shown in Table 3.5.

After calculating the fitness of solutions, they can be arranged according to the requirement of operators. The selection operator places solutions according to their fitness. We can sort them in ascending order according to the fitness value of each solution. Therefore, the solution with minimum fitness will be placed at the top, and the solution with maximum fitness will be stored last.

In addition to cost, their position should also arrange according to the fitness.

The following command can be used for this purpose:

`[cost, ind]=sort(cost)` // this command will arrange all solutions according to their fitness in ascending order with min-cost in element 1

TABLE 3.5
Initial Population along with Fitness Values

Population (par)	Cost
{0.2827, 1.0383, 2.3233}	93.5856
{-3.9976, -4.2840, 2.9327}	120.6962
{-0.5403, 4.5601, -2.9835}	117.8976
{3.8465, -3.7755, 2.4192}	109.1314
{4.4482, 4.0423, -4.1234}	135.6681
{1.9873, 1.2916, 4.3718}	114.4500
{3.7242, 1.2454, -2.8845}	101.9688
{4.5596, -3.6379, -2.7144}	108.1467
{-2.4269, -1.3617, 2.2370}	95.4700
{-3.9196, 4.1997, 3.3334}	142.0441

TABLE 3.6
Sorted Population (par) along with Fitness Values

Population	Cost
{0.2827, 1.0383, 2.3233}	93.5856
{-2.4269, -1.3617, 2.2370}	95.4700
{3.7242, 1.2454, -2.8845}	101.9688
{4.5596, -3.6379, -2.7144}	108.1467
{3.8465, -3.7755, 2.4192}	109.1314
{1.9873, 1.2916, 4.3718}	114.4500
{-0.5403, 4.5601, -2.9835}	117.8976
{-3.9976, -4.2840, 2.9327}	120.6962
{4.4482, 4.0423, -4.1234}	135.6681
{-3.9196, 4.1997, 3.3334}	142.0441

Sort cost values

```
cost = {93.5856, 95.4700, 101.9688, 108.1467, 109.1314,
114.4500, 117.8976, 120.6962, 135.6681, 142.0441}
```

Find indexes of vectors in ascending order of their costs

```
ind = {1,9, 7, 8, 4, 6, 3, 2, 5,10}
ind = {1 9 7 6 8 3 4 10 2 5}
```

This will place the min value at the top, and other values will be arranged accordingly, and their indices will be placed in the index value

After arranging all solutions, their positions can be ordered according to new indices. The following command will implement this:

```
par=par(ind,:);
```

Sort par matrix according to cost

```
par =
```

After arranging the population according to fitness now, we can apply the selection operator. Although several selection operators exist, how can we apply a selection operator yet? We can apply the same operator which we have used in the BGA. For generating an intermediate population, we first check how many solutions will take part in the selection.

3.3.2 Implementing Selection Operator in Real-Parameter GA Using MATLAB

After generating the initial population, the programmer will decide how many solutions will produce offspring. In some problems, only a few solutions in the population will participate in producing offspring. Some solutions will not take part in producing offspring. They are directly transferred from one generation to another generation. Such programs for GAs are called elitist GAs. Good solutions produced in previous generations are kept in a separate storage in such programs. These solutions are kept in separate storage because they might be lost in new generations if they produce offspring. Therefore, they are preserved for new generations.

The number of solutions that will participate in selection will be decided by the number of matings. These matings will depend on the selection probability and can be calculated with the following code:

```
ns=floor(selection*popsize)
M=ceil((popsize-ns)/2);
```

This will fix the value of the number of solutions (`ns`) at 5. After selection, the selected solutions will perform mating. These matings will be implemented during the crossover. M will be used to define the number of mating. In this program, the number of mating is taken as 3.

Mating will be implemented in pair, and in every mating, two solutions will be selected randomly from the population. Here roulette wheel selection is used for selecting solutions. Roulette wheel selection works on cumulative probability.

The following code will be used to define the selection.

For implementing roulette wheel selection, the portion of the fitness wheel covered by each solution can be determined by the following command:

`prob=flipud([1:ns]'/sum([1:ns]));` this will determine the proportion of each solution in creating the roulette wheel.

After checking the proportion of each solution, the portion of wheel can be allocated according to the fitness of the solutions.

odds = [0 cumsum(`prob(1:ns)`)'`];` *%* probability distribution function

Once these values are created, the solution participating in each mating can be determined by the following code.

The values of these variables in our program are as follows:

Number of matings

```
M = ceil((popsize-ns)/2) = upper value of ((10-5)/2)
sum([1:ns]) = [1 2 3 4 5]=15
```

Weights chromosomes

```
prob = flipud([1:ns]'/sum([1:ns])) = transpose of [5/15 4/15
3/15 2/15 1/15]
```

TABLE 3.7
Chromosome Weights

0.3333
0.2667
0.2000
0.1333
0.0667

TABLE 3.8
Table Showing Cumulative Odds

0
0.3333
0.6000
0.8000
0.9333
1.0000

Probability distribution function
Odds are calculated by adding the new value with the previous value. Starting with 0, the next value will be 0.3333. Similarly, the next value will be .3333+.2667=.6000. Similarly, other values can be calculated.

For the identification of the solution that will participate in mating, three random numbers are generated for the first solution and three random numbers are generated for the second solution.

Now for defining the index of solution, we check which solutions will be covering these random numbers. The code for selecting solutions for mating follows:

```
fix1=rand(1,M); % Mate #1
fix2=rand(1,M); % Mate #2
```

Mate 1
fix1 = rand(1,M) = {0.5999, 0.9544, 0.5273}

Mate 2
fix2 = rand(1,M) = {0.7466, 0.4118, 0.9289}

```
count=1;
while count<=M
for id=2:ns+1
if fix1(count)<=odds(id) & fix1(count)>odds(id-1)
ma(count)=id-1;
  end % if
if fix2(count)<=odds(id) & fix2(count)>odds(id-1)
pa(count) =id-1;
  end % if
  end % id
count=count+1;
end
```

3.3.3 Implementing Crossover in a Real-Parameter GA Using MATLAB

This code is used to select a solution for performing crossover. According to the code, the number of mating will be 3. In each mating, two solutions will participate in the crossover and mutation operations. The index values of these solutions are identified on the basic of cumulative fitness.

Two solutions which are selected for mating will be decided randomly on the basis of cumulative probability.

ma and pa contain the indices of the chromosomes that will mate

```
count = 1
ma = {2}
pa = {3}
count = 2
pa = {3,2}
ma = {2,5}
count = 3
ma = {2, 5, 2}
pa = {3, 2, 4}
```

3.3.3.1 Implementing a Crossover Operator

After selection crossover operator is performed to generate offspring solutions. A cut will determine the nature of offspring. If cut is near to least significant bit, then exploitation can be performed. If it is near to the most significant digit, then it will contribute to exploring the areas. The following code is used to apply the crossover.

% Performs mating using single point crossover

```
    ix=1:2:ns; % index of mate #1
    xp=ceil(rand(1,M)*Dim); % crossover point
    r=rand(1,M); % mixing parameter
for ic=1:M
    xy=par(ma(ic),xp(ic))-par(pa(ic),xp(ic)); % ma and pa
    % mate
    par(ns+ix(ic),:)=par(ma(ic),:); % 1st offspring
    par(ns+ix(ic)+1,:)=par(pa(ic),:); % 2nd offspring
par(ns+ix(ic),xp(ic))=par(ma(ic),xp(ic))-r(ic).*xy;
    % 1st
par(ns+ix(ic)+1,xp(ic))=par(pa(ic),xp(ic))+r(ic).*xy;
    % 2nd
    If xp(ic)<Dim % crossover when last variable not selected
par(ns+ix(ic),:)=[par(ns+ix(ic),1:xp(ic)),par(ns+ix(ic)+1,
xp(ic)+1:Dim)];
par(ns+ix(ic)+1,:)=[par(ns+ix(ic)+1,1:xp(ic)),par(ns+ix(ic),
xp(ic)+1:Dim)];
    end % if
end
```

When executed, the following values will be generated:

Index of mate 1

```
ix = {1, 3, 5}
```

Crossover point

```
xp = {1, 1, 1}
```

Mixing parameter

```
r = {0.9947, 0.7425, 0.5483}
```

Mating 1 of ma and pa

```
xy = par(2,1)-par(3,1)= -2.4269-3.7242
 -6.1511
```

The first offspring will be created by copying the second row of par to the sixth row of par.

```
par =
```

TABLE 3.9
First Offspring

```
{0.2827, 1.0383, 2.3233}
{-2.4269, -1.3617, 2.2370}
{3.7242, 1.2454, -2.8845}
{4.5596, -3.6379, -2.7144}
{3.8465, -3.7755, 2.4192}
{-2.4269, -1.3617, 2.2370}
{-0.5403, 4.5601, -2.9835}
{-3.9976, -4.2840, 2.9327}
{4.4482, 4.0423, -4.1234}
{-3.9196, 4.1997, 3.3334}
```

Note: The black and bold value represents the data which have been copied.

The second offspring, similarly, will be created by copying the third row to the seventh row. Modified par matrix can be shown as:

```
par =
```

TABLE 3.10
Second Offspring

```
{0.2827, 1.0383, 2.3233}
{-2.4269, -1.3617, 2.2370}
{3.7242, 1.2454, -2.8845}
{4.5596, -3.6379, -2.7144}
{3.8465, -3.7755, 2.4192}
{-2.4269, -1.3617, 2.2370}
{3.7242, 1.2454, -2.8845}
{-3.9976, -4.2840, 2.9327}
{4.4482, 4.0423, -4.1234}
{-3.9196, 4.1997, 3.3334}
```

Note: The black and bold value represents the data which have been copied.

After copying these rows, some members of these are modified according to the selected index of crossover. Changes in values are reflected by black in Table 3.11.

Crossover according to crossover point

```
par =
```

TABLE 3.11
Crossover Point Selection

{0.2827, 1.0383, 2.3233}
{-2.4269, -1.3617, 2.2370}
{3.7242, 1.2454, -2.8845}
{4.5596, -3.6379, -2.7144}
{3.8465, -3.7755, 2.4192}
{**3.6915**, -1.3617, 2.2370}
{3.7242, 1.2454, -2.8845}
{-3.9976, -4.2840, 2.9327}
{4.4482, 4.0423, -4.1234}
{-3.9196, 4.1997, 3.3334}

Note: The black and bold value represents the data which have been copied.

Similarly, the other value can be changed. These random changes will change the matrix to this modified value.

```
par =
```

TABLE 3.12
Population after Crossover

{0.2827, 1.0383, 2.3233}
{-2.4269, -1.3617, 2.2370}
{3.7242, 1.2454, -2.8845}
{4.5596, -3.6379, -2.7144}
{3.8465, -3.7755, 2.4192}
{3.6915, -1.3617, 2.2370}
{-2.3942, 1.2454, -2.8845}
{-3.9976, -4.2840, 2.9327}
{4.4482, 4.0423, -4.1234}
{-3.9196, 4.1997, 3.3334}

Crossover when last variable not selected

```
par =
```

TABLE 3.13
Crossover When Last Variable Is Not Selected

{0.2827, 1.0383, 2.3233}
{-2.4269, -1.3617, 2.2370}
{3.7242, 1.2454, -2.8845}
{4.5596, -3.6379, -2.7144}
{3.8465, -3.7755, 2.4192}
{3.6915, 1.2454, -2.8845}
{-2.3942, 1.2454, -2.8845}
{-3.9976, -4.2840, 2.9327}
{4.4482, 4.0423, -4.1234}
{-3.9196, 4.1997, 3.3334}

Note: The black and bold value represents the data which have been copied.

par =

TABLE 3.14
Modified Population

{0.2827, 1.0383, 2.3233}
{-2.4269, -1.3617, 2.2370}
{3.7242, 1.2454, -2.8845}
{4.5596, -3.6379, -2.7144}
{3.8465, -3.7755, 2.4192}
{3.6915, 1.2454, -2.8845}
{-2.3942, 1.2454, -2.8845}
{-3.9976, -4.2840, 2.9327}
{4.4482, 4.0423, -4.1234}
{-3.9196, 4.1997, 3.3334}

Note: The black and bold value represents the data which have been copied.

Mating 2 of ma and pa

Similarly updated values for other matings can be calculated.

xy = 6.2733

First offspring

par =

TABLE 3.15
First Offspring

{0.2827, 1.0383, 2.3233}
{-2.4269, -1.3617, 2.2370}
{3.7242, 1.2454, -2.8845}
{4.5596, -3.6379, -2.7144}
{3.8465, -3.7755, 2.4192}
{3.6915, 1.2454, -2.8845}
{-2.3942, 1.2454, -2.8845}
{3.8465, -3.7755, 2.4192}
{4.4482, 4.0423, -4.1234}
{-3.9196, 4.1997, 3.3334}

Note: The black and bold value represents the data which have been copied.

Second offspring

par =

TABLE 3.16
Second Offspring

{0.2827, 1.0383, 2.3233}
{-2.4269, -1.3617, 2.2370}
{3.7242, 1.2454, -2.8845}
{4.5596, -3.6379, -2.7144}
{3.8465, -3.7755, 2.4192}
{3.6915, 1.2454, -2.8845}
{-2.3942, 1.2454, -2.8845}
{3.8465, -3.7755, 2.4192}
{-2.4269, -1.3617, 2.2370}
{-3.9196, 4.1997, 3.3334}

Note: The black and bold value represents the data which have been copied.

Crossover according to crossover point

par =

TABLE 3.17
Crossover Point Selection

{0.2827, 1.0383, 2.3233}
{-2.4269, -1.3617, 2.2370}
{3.7242, 1.2454, -2.8845}
{4.5596, -3.6379, -2.7144}

TABLE 3.17 (Continued))

{3.8465, -3.7755, 2.4192}
{3.6915, 1.2454, -2.8845}
{-2.3942, 1.2454, -2.8845}
{-0.8113, -3.7755, 2.4192}
{-2.4269, -1.3617, 2.2370}
{-3.9196, 4.1997, 3.3334}

Note: The black and bold value represents the data which have been copied.

```
par =
```

TABLE 3.18
Modified Population

{0.2827, 1.0383, 2.3233}
{-2.4269, -1.3617, 2.2370}
{3.7242, 1.2454, -2.8845}
{4.5596, -3.6379, -2.7144}
{3.8465, -3.7755, 2.4192}
{3.6915, 1.2454, -2.8845}
{-2.3942, 1.2454, -2.8845}
{-0.8113, -3.7755, 2.4192}
{2.2309, -1.3617, 2.2370}
{-3.9196, 4.1997, 3.3334}

Note: The black and bold value represents the data which have been copied.

<u>Crossover when last variable not selected</u>

```
par =
```

TABLE 3.19
When Last Variable Is Not Selected for Crossover

{0.2827, 1.0383, 2.3233}
{-2.4269, -1.3617, 2.2370}
{3.7242, 1.2454, -2.8845}
{4.5596, -3.6379, -2.7144}
{3.8465, -3.7755, 2.4192}
{3.6915, 1.2454, -2.8845}
{-2.3942, 1.2454, -2.8845}
{-0.8113, -1.3617, 2.2370}
{2.2309, -1.3617, 2.2370}
{-3.9196, 4.1997, 3.3334}

Note: The black and bold value represents the data which have been copied.

```
par =
```

TABLE 3.20
Modified Population

{0.2827, 1.0383, 2.3233}
{-2.4269, -1.3617, 2.2370}
{3.7242, 1.2454, -2.8845}
{4.5596, -3.6379, -2.7144}
{3.8465, -3.7755, 2.4192}
{3.6915, 1.2454, -2.8845}
{-2.3942, 1.2454, -2.8845}
{-0.8113, -1.3617, 2.2370}
{2.2309, -1.3617, 2.2370}
{-3.9196, 4.1997, 3.3334}

Note: The black and bold value represents the data which have been copied.

Mating 3 of ma and pa

When a third mating is used, then the intermediate population size will be increased to 11. This updated population size will also be used during mutation. However, after completing selection, crossover, and mutation, only the 10 best solutions will be kept for the next generation.

```
xy = -6.9865
```

First offspring

```
par =
```

TABLE 3.21
First Offspring

{0.2827, 1.0383, 2.3233}
{-2.4269, -1.3617, 2.2370}
{3.7242, 1.2454, -2.8845}
{4.5596, -3.6379, -2.7144}
{3.8465, -3.7755, 2.4192}
{3.6915, 1.2454, -2.8845}
{-2.3942, 1.2454, -2.8845}
{-0.8113, -1.3617, 2.2370}
{2.2309, -1.3617, 2.2370}
{-2.4269, -1.3617, 2.2370}

Note: The black and bold value represents the data which have been copied.

Second offspring

```
par =
```

TABLE 3.22
Second Offspring

{0.2827, 1.0383, 2.3233}
{-2.4269, -1.3617, 2.2370}
{3.7242, 1.2454, -2.8845}
{4.5596, -3.6379, -2.7144}
{3.8465, -3.7755, 2.4192}
{3.6915, 1.2454, -2.8845}
{-2.3942, 1.2454, -2.8845}
{-0.8113, -1.3617, 2.2370}
{2.2309, -1.3617, 2.2370}
{-2.4269, -1.3617, 2.2370}
{4.5596, -3.6379, -2.7144}

Note: The black and bold value represents the data which have been copied.

Crossover according to crossover point

```
par =
```

TABLE 3.23
Crossover Point Selection

{0.2827, 1.0383, 2.3233}
{-2.4269, -1.3617, 2.2370}
{3.7242, 1.2454, -2.8845}
{4.5596, -3.6379, -2.7144}
{3.8465, -3.7755, 2.4192}
{3.6915, 1.2454, -2.8845}
{-2.3942, 1.2454, -2.8845}
{-0.8113, -1.3617, 2.2370}
{2.2309, -1.3617, 2.2370}
{1.4035, -1.3617, 2.2370}
{4.5596, -3.6379, -2.7144}

Note: The black and bold value represents the data which have been copied.

```
par =
```

TABLE 3.24
Modified Population

{0.2827, 1.0383, 2.3233}
{-2.4269, -1.3617, 2.2370}
{3.7242, 1.2454, -2.8845}
{4.5596, -3.6379, -2.7144}
{3.8465, -3.7755, 2.4192}
{3.6915, 1.2454, -2.8845}
{-2.3942, 1.2454, -2.8845}
{-0.8113, -1.3617, 2.2370}
{2.2309, -1.3617, 2.2370}
{1.4035, -1.3617, 2.2370}
{**0.7293**, -3.6379, -2.7144}

Note: The black and bold value represents the data which have been copied.

Crossover when last variable not selected

```
par =
```

TABLE 3.25
Crossover When Last Variable Is Not Selected

{0.2827, 1.0383, 2.3233}
{-2.4269, -1.3617, 2.2370}
{3.7242, 1.2454, -2.8845}
{4.5596, -3.6379, -2.7144}
{3.8465, -3.7755, 2.4192}
{3.6915, 1.2454, -2.8845}
{-2.3942, 1.2454, -2.8845}
{-0.8113, -1.3617, 2.2370}
{2.2309, -1.3617, 2.2370}
{1.4035, -3.6379, -2.7144}
{0.7293, -3.6379, -2.7144}

Note: The black and bold value represents the data which have been copied.

```
par =
```

TABLE 3.26
Modified Population

{0.2827, 1.0383, 2.3233}
{-2.4269, -1.3617, 2.2370}
{3.7242, 1.2454, -2.8845}
{4.5596, -3.6379, -2.7144}

TABLE 3.26 (Continued))

{3.8465, -3.7755, 2.4192}
{3.6915, 1.2454, -2.8845}
{-2.3942, 1.2454, -2.8845}
{-0.8113, -1.3617, 2.2370}
{2.2309, -1.3617, 2.2370}
{1.4035, -3.6379, -2.7144}
{0.7293, -3.6379, -2.7144}

Note: The black and bold value represents the data which have been copied.

3.3.4 Mutation Operation

After crossover, mutation will be applied to generate the final offspring.
Following is the code for implementing mutation.

% Mutate the population

```
mrow=sort(ceil(rand(1,nmut)*(popsize-1))+1);
mcol=ceil(rand(1,nmut)*Dim);
for ii=1:nmut

    par(mrow(ii),mcol(ii))=(varhi-varlo)*rand+varlo; % mutation

End
```

When this code is executed, the following values will be generated. The first two lines of code are used to generate the row and column numbers of the index to be mutated.

```
t1 = rand(1,nmut) =
{0.6867,0.6116,0.8877,0.4945,0.6276,0.8116}
mrow = sort(ceil(t1*(popsize-1))+1) = {6, 7, 7, 8, 9, 9}
t2 = rand(1,nmut) =
{0.3728,0.6364,0.2160,0.8858,0.9327,0.9607}
mcol = ceil(t2*Dim) = {2,2,1,3,3,3}
```

After generating the index of each solution, these indexes are replaced by completely new values. The following is an example.

```
ii = 1
t3 = rand = 0.6304
par(mrow(ii),mcol(ii)) = par(mrow(1),mcol(1)) = par(6,2) =
(varhi-varlo)*t3+varlo
par =
```

TABLE 3.27
Mutation 1

```
{0.2827, 1.0383, 2.3233}
{-2.4269, -1.3617, 2.2370}
{3.7242, 1.2454, -2.8845}
{4.5596, -3.6379, -2.7144}
{3.8465, -3.7755, 2.4192}
{3.6915, 1.3043, -2.8845}
{-2.3942, 1.2454, -2.8845}
{-0.8113, -1.3617, 2.2370}
{2.2309, -1.3617, 2.2370}
{1.4035, -3.6379, -2.7144}
{0.7293, -3.6379, -2.7144}
```

Note: The black and bold value represents the data which have been copied.

Similarly, other values can be changed.

```
ii = 2
t3 = rand = 0.1588
par(mrow(ii),mcol(ii)) = par(mrow(2),mcol(2)) = par(7,2) =
(varhi-varlo)*t3+varlo
par =
```

TABLE 3.28
Mutation 2

```
{0.2827, 1.0383, 2.3233}
{-2.4269, -1.3617, 2.2370}
{3.7242, 1.2454, -2.8845}
{4.5596, -3.6379, -2.7144}
{3.8465, -3.7755, 2.4192}
{3.6915, 1.3043, -2.8845}
{-2.3942, -3.4121, -2.8845}
{-0.8113, -1.3617, 2.2370}
{2.2309, -1.3617, 2.2370}
{1.4035, -3.6379, -2.7144}
{0.7293, -3.6379, -2.7144}
```

Note: The black and bold value represents the data which have been copied.

This process is repeated until all values are changed and the mutation operation is completed.

```
ii = 3
t3 = rand = 0.9481
par(mrow(ii),mcol(ii)) = par(mrow(3),mcol(3)) = par(7,1) =
(varhi-varlo)*t3+varlo
```

```
par =
```

TABLE 3.29
Mutation 3

{0.2827, 1.0383, 2.3233}
{-2.4269, -1.3617, 2.2370}
{3.7242, 1.2454, -2.8845}
{4.5596, -3.6379, -2.7144}
{3.8465, -3.7755, 2.4192}
{3.6915, 1.3043, -2.8845}
{**4.4805,** -3.4121, -2.8845}
{-0.8113, -1.3617, 2.2370}
{2.2309, -1.3617, 2.2370}
{1.4035, -3.6379, -2.7144}
{0.7293, -3.6379, -2.7144}

Note: The black and bold value represents the data which have been copied.

```
ii = 4
t3 = rand = 0.7056
par(mrow(ii),mcol(ii)) = par(mrow(4),mcol(4)) = par(8,3) =
(varhi-varlo)*t3+varlo
par =
```

TABLE 3.30
Mutation 4

{0.2827, 1.0383, 2.3233}
{-2.4269, -1.3617, 2.2370}
{3.7242, 1.2454, -2.8845}
{4.5596, -3.6379, -2.7144}
{3.8465, -3.7755, 2.4192}
{3.6915, 1.3043, -2.8845}
{4.4805, -3.4121, -2.8845}
{-0.8113, -1.3617, **2.0562**}
{2.2309, -1.3617, 2.2370}
{1.4035, -3.6379, -2.7144}
{0.7293, -3.6379, -2.7144}

Note: The black and bold value represents the data which have been copied.

```
ii = 5
t3 = rand = 0.2029
par(mrow(ii),mcol(ii)) = par(mrow(5),mcol(5)) = par(9,3) =
(varhi-varlo)*t3+varlo
par =
```

TABLE 3.31
Mutation 5

```
{0.2827, 1.0383, 2.3233}
{-2.4269, -1.3617, 2.2370}
{3.7242, 1.2454, -2.8845}
{4.5596, -3.6379, -2.7144}
{3.8465, -3.7755, 2.4192}
{3.6915, 1.3043, -2.8845}
{4.4805, -3.4121, -2.8845}
{-0.8113, -1.3617, 2.0562}
{2.2309, -1.3617, -2.9708}
{1.4035, -3.6379, -2.7144}
{0.7293, -3.6379, -2.7144}
```

Note: The black and bold value represents the data which have been copied.

```
ii = 6
t3 = rand = 0.0373
par(mrow(ii),mcol(ii)) = par(mrow(6),mcol(6)) = par(9,3) =
(varhi-varlo)*t3+varlo
par =
```

TABLE 3.32
Mutation 6

```
{0.2827, 1.0383, 2.3233}
{-2.4269, -1.3617, 2.2370}
{3.7242, 1.2454, -2.8845}
{4.5596, -3.6379, -2.7144}
{3.8465, -3.7755, 2.4192}
{3.6915, 1.3043, -2.8845}
{4.4805, -3.4121, -2.8845}
{-0.8113, -1.3617, 2.0562}
{2.2309, -1.3617, -4.6269}
{1.4035, -3.6379, -2.7144}
{0.7293, -3.6379, -2.7144}
```

Note: The black and bold value represents the data which have been copied.

Completion of the first iteration and the production of new solutions:

After selection, crossover, and mutation, a new population is generated and re-evaluated, as shown in Table 3.33.

TABLE 3.33
New Population with Fitness Value

Population	Cost
{0.2827, 1.0383, 2.3233}	93.5856
{-2.4269, -1.3617, 2.2370}	95.4700
{3.7242, 1.2454, -2.8845}	101.9688
{4.5596, -3.6379, -2.7144}	108.1467
{3.8465, -3.7755, 2.4192}	109.1314
{3.6915, 1.3043, -2.8845}	102.0298
{4.4805, -3.4121, -2.8845}	107.1077
{-0.8113, -1.3617, 2.0562}	88.3841
{2.2309, -1.3617, -4.6269}	98.6674
{1.4035, -3.6379, -2.7144}	90.9220
{0.7293, -3.6379, -2.7144}	89.8249

You can see the updated population contains 11 solutions while the size used for a GA was 10. During selection, the 10 best solutions from these solutions will participate in producing the next generation.

4 Differential Evolution

In previous chapters, we discussed various versions of genetic algorithms (GAs). We also discussed how each version of the GA was created by mapping natural selection and evolution theory. The overall process of natural selection and evolution has been implemented in the GA using the three operators: selection, crossover, and mutation. Several questions arise after reading the GA theory. For example, is it the only algorithm based on the concepts of natural selection and evolution? Is it necessary to use the selection operator first and then proceed to reproduction (which includes crossover and mutation)? Are the equations used in the various operators of the GA fixed, or can they be changed? Can we change the equations, and how do we do it?

The following sections discuss the answers to these questions and discuss a new evolutionary algorithm known as the differential evolution (DE) algorithm. We also explain why the equations used in the DE algorithm will help solve optimization problems.

4.1 INTRODUCTION

In addition to the GA, several other algorithms such as DE, evolutionary strategy, and evolutionary programming have been developed by mapping the theory of survival of the fittest. DE, like the GA, is a popular algorithm. DE and GA use three operators—selection, crossover, and mutation—to implement the same phenomenon of natural selection and evolution. However, the order in which the operators are implemented differs in both algorithms. In a GA, the order is selection, crossover, and mutation, whereas in DE, the order is mutation, crossover, and selection.

They were also created to solve different types of optimization problems. While DE is used in numerical optimization problems to minimize the fitness function, a GA was designed to solve discrete problems and maximize the objective function's value. In contrast to a GA and DE, evolutionary strategy and evolutionary programming use only two operators, selection and mutation, to put the theory into action. The concept of crossover does not exist. The Environmental Adaption Method (EAM) is founded on adaptive learning theory, a subprocess of evolutionary learning. EAM implements the program with the help of three operators: adaptation, mutation, and selection. For the time being, let us focus on the DE algorithm.

4.2 DE ALGORITHMS

DE was proposed in 1995 by Rainer Storn and Kenneth Price. It is used to solve numerical optimization problems in which we want to minimize the value of the objective function. In general, DE solves continuous optimization problems with real-valued variables. This algorithm uses mutation, crossover, and selection operators to update the parent population's solution. One iteration is completed after

DOI: 10.1201/9781003313649-4

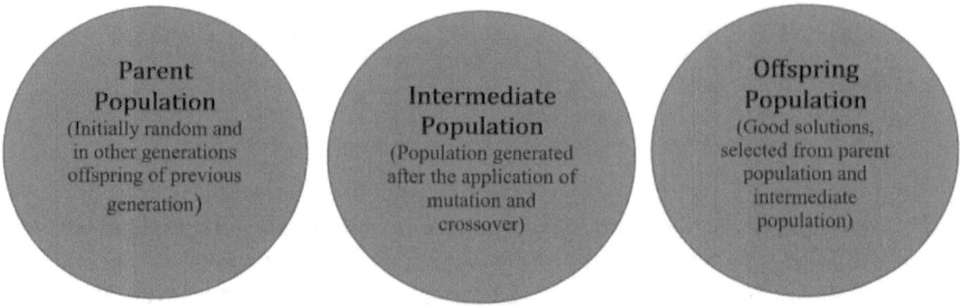

FIGURE 4.1 DE explained.

conversion from the parent population to the offspring population. This process is repeated until the solution with the desired fitness is obtained, or the number of iterations equals the maximum number of iterations. A generation defines a set of solutions present at the same time. Parent solutions are generated from the entire search space in the initial generation. In later generations, the offspring population of the previous generation will act as the parent population. Later on, these solutions are converged to the optimum solution of the problem by applying different operators. Graphically, the process of DE can be explained by Figure 4.1.

To solve the problem, we must first determine the total number of iterations (maximum iterations). An experienced programmer estimates the maximum number of iterations based on the complexity of the problem. We will convert the parent population to the offspring population in one iteration. We use a counter to count iterations, and once the counter reaches the maximum number of iterations, we stop. We halt the algorithm. Each iteration converts the parent population into offspring using mutation, crossover, and selection operators.

4.2.1 Mutation Operator

According to the terminology used in the literature created for DE algorithms, the solutions of the parent population are referred to as target vectors. The first operator in DE is mutation. It is applied to every solution in the parent population, and mutant solutions are created. The solutions produced due to mutation are referred to as donor vectors. In general, the mutation operator's purpose is to provide variety among solutions in initial iterations, and in later iterations, they are used to exploit the regions around good solutions. In later iterations mutation, mutated solutions will be very different in the first generation. However, as the number of iterations increases, they converge on points that may solve the problem. In the past, several mutation strategies have been used for defining the mutation operation.

Let us take an example and discuss how mutation can be implemented. Let us suppose the size of the initial population is NP and r_1, r_2, *and* r_3 are the value of indices of parent vectors in gth generation. A commonly used mutation operator follows:

Mutation: During mutation, a donor vector $\vec{v}$ is generated for each target vector $\vec{X}_{i,G}$ in the current population using the mutation strategy; that is,

$$\vec{V}_{i,G} = \vec{X}_{r_1,G} + F.\left(\vec{X}_{r_2,G} - \vec{X}_{r_3,G}\right), \quad where\ r_1 \neq r_2 \neq r_3 \neq i, \tag{4.1}$$

with randomly chosen indexes $r_1, r_2, and\, r_3 \in [1, NP]$. It is clear from the equation that the solution for which mutant is created will not participate in producing mutant solution. Each parent vector/target vector will choose three random solutions from the current population for generating the donor vector. After that, the difference of two random vectors is added to the third random vector to produce a donor vector. F is a scaling factor $F \in [0,2]$, which controls the amplification of the difference vector $\left(\vec{X}_{r_2,G} - \vec{X}_{r_3,G}\right)$. The role of mutation is to perform exploration by maintaining diversity among solutions.

4.2.2 Crossover Operator

As discussed earlier, that mutation operator is used to generate diverse solutions. However, only developing diverse solutions is not enough. They should be combined with good solutions to exploit good areas in the search space. This is implemented in DE by crossover and selection operator. After mutation, each target vector $\vec{X}_{ji,G}$ is combined with its corresponding mutant vector $\vec{V}_{ji,G+1}$ to generate trial vector $\vec{U}_{ji,G+1}$. This can be implemented by using two crossover operators, that is, binomial and exponential crossovers.

The binomial crossover is as follows:

$$\vec{u}_{i,G+1} = \left(\vec{u}_{i,G+1}, \vec{u}_{2i,G+1}, \ldots, \vec{u}_{Di,G+1}\right), \tag{4.2}$$

where

$$\vec{u}_{ji,G+1} = \{\vec{v}_{ji,G+1}\ if\ r(j) \leq CR\ or\ j = rn(i), \vec{x}_{ji,G}\ if\ r(j) > CR\ and\ j \neq rn(i), \tag{4.3}$$

where $j = 1,2,\ldots,D$ $r(j) \in [0,1]$ is the jth evaluation of a uniform random generator number. CR is the crossover constant $[0,1], rn(i)$ $(1,2,\ldots D)$ is a randomly chosen index to ensure $\vec{u}_{ji,G+1}$ gets at least one element from $\vec{v}_{ji,G+1}$. Otherwise, no new parent vector would be produced, and then the generated population would not be altered.

4.2.3 Selection Operator

Selection is used to decide whether the newly generated trail vector is preferred over the target vector. The selection operator is based on the theory of survival of the fittest. The fitness of the trial and target vector is compared, and a better solution with better fitness is selected for the next generation. The selection operator for minimization problems is expressed as follows:

$$\vec{x}_{i,G+1} = \{\vec{u}_{i,G+1}, if \;\; f(\vec{u}_{i,G+1}) < f(\vec{x}_{i,G}),\, \vec{x}_{i,G} \;\; otherwise \tag{4.4}$$

For $j=1,2,..,D$. If the trial vector $\vec{u}_{i,G+1}$ gives a better fitness value than $\vec{x}_{i,G}$, then $\vec{u}_{i,G+1}$ is set to $\vec{x}_{i,G+1}$; otherwise, the old value of $\vec{x}_{i,G}$ is continued.

4.2.4 Control Parameters

Some parameters can change the performance of DE algorithm. These parameters are called control parameters, which control the convergence rate of DE algorithm. These parameters are discussed next.

Mutation Scaling Factor (F): The mutation scaling factor is used to increase the exploration capability of the algorithm. It is used to generate solutions from the entire search space so that local optima can escape and a global optimum solution can be captured. In most versions of DE, it was taken as 0.8. However, we can choose any value between (0, 2). As the value of F increases, the diversity among solutions increases, and the convergence rate is decreased. In some variants of DE, these control parameters are changed automatically as per requirement. These versions of DE are called self-adaptive variants of DE.

Crossover Rate (Cr): Crossover rate is another parameter that affects the DE algorithm's convergence rate. Low values of crossover probability *Cr* are used to generate a new solution close to the target vector. However, higher values of *Cr* are used to provide diversity among solutions.

Low values of *Cr*-like values (0 to 0.2) will perform significantly fewer changes in the target vectors. On the other hand, high values of *Cr* (i.e., 0.9 to 1.0) are used for multimodal problems to escape local optima. It also speeds up the convergence with an increased convergence rate.

Population Size (NP): In most problems, 40 solutions will be sufficient in population. However, to solve higher dimension problems, say, 40 * D, it may increase performance (where D is the dimension).

Algorithm 4.1: Classical DE Algorithm

1: **procedure** DE
2: generate random vectors $\vec{x}_{i,G}$ of population
3: **for** $i=1$ to maximum iterations **do**
4: **for** each vector $\vec{x}_{i,G}$ of population N **do**
5: MUTATION

::::::::::::::::::::
select three random vectors for mutation
$\vec{v}_{i,G} = \vec{x}_{r_1,G} + F * (\vec{x}_{r_2,G} - \vec{x}_{r_3,G})$
CROSSOVER
:::::::::::::::::::::
if $r(j) \leq CR \, or \, j = rn(i)$ **then**
$\vec{u}_{ji,G+1} = \vec{v}_{ji,G+1}$
else
$\vec{v}_{ji,G+1} = \vec{x}_{ji,G}$
SELECTION
:::::::::::::::::::::
evaluate fitness of $\vec{x}_{i,G}$ and $\vec{u}_{i,G+1}$ and store best solution in $\vec{x}_{i,G+1}$
replace $\vec{x}_{i,G}$ with $\vec{x}_{i,G+1}$ and repeat the process
end for
end for
end procedure

4.3 MORE ON DE ALGORITHMS

DE is a straightforward algorithm that is very simple to implement in a computer program. Because of its simplicity quickly became a popular tool for solving optimization problems, particularly in the continuous search space. It has been used to solve a variety of issues. Although DE and GAs appear to be very similar, they use the same operators to generate new solutions, namely, mutation, crossover, and selection, yet in GAs, the crossover is the primary operator because it is given precedence over mutation. In DE, on the other hand, the mutation is the primary operation and crossover is given less importance. DE is very effective compared to binary GAs in solving real-valued problems as it does not require conversion from binary value to real value. In terms of solution accuracy and convergence speed, DE outperformed many algorithms in competitions organized by top-level conferences.

Its performance is excellent, and like GAs, it has been applied to solve many numerical single-objective optimization problems. As DE works with real-parameter encoding, it encodes solutions as real-valued vectors, updated to construct new solutions using vector addition, scalar multiplication, and exchange of components (crossover). After creating a new solution, each solution is compared with its parent. If the candidate is better than its parent, it replaces the parent in the population. Otherwise, the candidate is discarded. As a steady-state algorithm, DE implicitly incorporates elitism; that is, no solution can be deleted from the population unless a better solution is found. While being a very successful optimization method, DE's greatest limitation originates in its encoding. As no vector solution representation exists for combinatorial problems, DE can only be applied in numerical optimization.

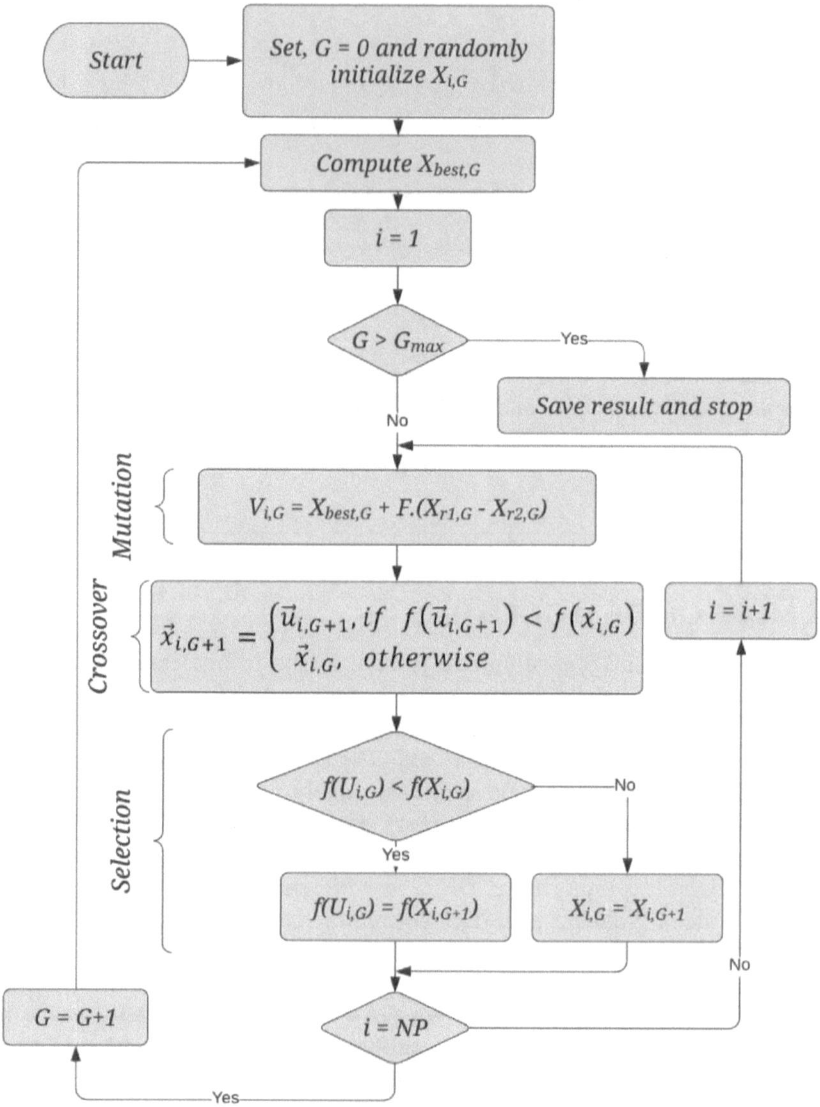

FIGURE 4.2 Flowchart of a DE algorithm.

TABLE 4.1
Notations Used in DE Algorithms

Notations	Description
POP	Current population
CR	Crossover Probability
F	Differential weight

TABLE 4.1 Continued

Notations	Description
Mgen	Maximum number of generations
final_POP	Newly generated population
final_fitness	Fitness of newly generated population
ps	Population size
Dim	Dimension

Algorithm 4.3: Algorithm of DE

```
DE()
begin
POP = rand(ps,Dim) // Initialize the population
Set the value of crossover probability CR where CR ∈ [0,1]
Set the value of differential weight F where F ∈ [0,2]
evaluate the function cost for POP
do
for each index = 1 to Mgen do
for each individual X in POP do //start loop through population
Select 3 unique random integers a, b and c in the range of 1 to POP where
a, b and c values are not equal to position of X
Select a number D in the range of 1 to Dim // select a dimension randomly
for each j = 1 to Dim do
Generate a random number k
//generate trial vector which is the outcome of binary crossover of POP agent
with intermediate agent (z =c+F*(a-b))
if((k < CR) or (i == Dim))
trial _ vector(index,D) = POP(c,D) + F*(POP(a,D)
- POP(b,D))
else
trial _ vector(index,D) = POP(index,D)
endif
D = (D+1)%Dim//get next dimension
end
evaluate the function cost for trial vector
if(cost(trial _ vector) <= cost(X)) // see if newly generated solu-
tion is better than original
final _ POP(index) = trial _ vector
final _ fitness(index) = cost(trial _ vector)
else
```

```
final _ POP(index) = X
final _ fitness(index) = cost(trial _ vector)
endif
 end
POP = final _ POP //update population until stopping criterion is met
end
```

4.4 MATLAB® IMPLEMENTATION OF A DE ALGORITHM

Let us take an example and explain the different steps of the algorithm. This program is written in MATLAB. Following is a step-by-step explanation of the algorithm.

4.4.1 Defining the Initial Population

First, we need to define the size of the initial population, and then, we can generate the initial population. Since DE works on real-parameter values, we need to generate real-parameter solutions. These real-parameter solutions must be generated within the given lower and upper bounds. Let us define the size of the initial population as 10 solutions for our problem. A `popsize` variable in MATLAB can be used to initialize this. We can initialize the population by setting `popsize` to `10`. So, for population initialization, we can use the statement `popsize = 10`.

After determining the size of the population, we need to generate 10 solutions in the search domain defined by lower and upper bounds. These bounds are represented by xmin and xmax. In addition to these bounds, we also define a boundary variable, xbound, which is 5 in this case.

```
So all important parameters related to DE can be defined at one
place
popsize = 10; (population size)
F = 1;    (for performing mutation)
CR = 0.5;    (crossover probability)
xbound = 5;   (common boundry value)
xmin = {-5, -5, -5};   (lower bound)
xmax = {5, 5, 5};   (upper bound)
```

Since the lower bound of search space is defined by xmin, the upper bound is defined by xmax. This indicates that any vector that is generated between the lower and upper bounds (including both boundary vector) will be included in the search space.

Following commands can be used to initialize solutions.

First, generate random values between `0` to `1` by function; this can be implemented by defining

```
rand(popsize,DIM) function.
t1 = rand(popsize,DIM) =
```

TABLE 4.2
Generating Solutions between 0 to 1

{0.7000, 0.5825, 0.9379}
{0.9174, 0.6473, 0.1218}
{0.7008, 0.8970, 0.1386}
{0.6835, 0.5561, 0.4567}
{0.3710, 0.6643, 0.2020}
{0.1566, 0.2730, 0.1634}
{0.0455, 0.7417, 0.8677}
{0.5407, 0.4070, 0.3448}
{0.6293, 0.1609, 0.8267}
{0.5091, 0.4316, 0.8392}

TABLE 4.3
Initial Population with Fitness

Population	Function Cost
{1.9998, 0.8246, 4.3788}	112.4972
{4.1743, 1.4728,-3.7822}	111.1255
{2.0076, 3.9700,-3.6144}	117.1981
{1.8354, 0.5612,-0.4329}	85.0209
{-1.2904, 1.6430,-2.9800}	94.7899
{-3.4342,-2.2697,-3.3660}	101.2922
{-4.5445, 2.4167, 3.6767}	134.6306
{0.4069,-0.9304,-1.5523}	80.2410
{1.2927,-3.3910, 3.2671}	101.4818
{0.0910,-0.6836, 3.3921}	96.6725

Then initial population will be generated by the equation

```
oldpop = 2 * xbound * rand(popsize,DIM) - xbound
  = 2 * xbound * t1 - xbound
```

After generating initial values fitness values of these solutions will be calculated. The initial population along with fitness value is shown in Table 4.3.

Another method for generating 10 random solutions follows. This method is used in those cases where the min and max bounds are different.

```
Oldpop=xmin+rand(popsize,DIM).*(xmax-xmin);
```

4.4.2 Updating Solutions by Using Mutation and a Crossover Operator

After generating the initial population, the mutation operator is used to generate new solutions. During mutation, each solution is replaced by a mutant vector.

Three solutions from the population are chosen (excluding the solution for which the mutant vector is calculated) to create the mutation vector for each solution. After selecting these three random vectors, the difference between the two vectors is added to the remaining vector, and the mutant vector is calculated.

After creating the donor vector, it is combined with previous vector values to generate the trial vector.

Let us look at how the entire process is carried out. This example is for the first solution. Other solutions can be generated in the same manner.

As an example, let us create a mutant vector for the first solution. For that, we will generate three index positions excluding 1.

In DE, both the mutation and crossover operations can be applied jointly. The offspring population is generated after mutation and crossover. The conversion from parent vector to offspring vector is done dimension by dimension. Let us look at the process. Since the dimension of the solution vector is 3, in classical mutation, three solutions are chosen from the population. First we have to generate three unique vectors.

Generate three unique random numbers

```
num_index = 8, 9, 7
num_a = 8
num_b = 9
num_c = 7
```

Now for each dimension, the following steps are applied.

Generate a random dimension

```
trail_selection_index = randi(DIM) = 1
```

For particle 1 and first dimension:

```
(perform crossover)
Generate a random number for crossover operation:
trail_rand_num = 0.7349
CR = 0.5
As trail_rand_num> CR:
The parent part will be supplied as offspring.
Trail_num (particleindex, trail_selection_index) = trail_num
(1, 1) = oldpop(1,1) = 1.9998
```

For particle 1 and the second dimension:

(crossover operator)

```
trail_rand_num = 0.8296
CR = 0.5
As trail_rand_num> CR:
trail_num = (particleindex, trail_selection_index) = trail_num
(1, 2) = oldpop(1,2) = 0.8246
```

TABLE 4.4
New Population by Applying Mutation and Crossover

Oldpop	Trail_rand_num	Trail Vector
{1.9998,0.8246, 4.3788}	{0.7349,0.8296,0.2019}	{1.9998,0.8246, -1.1426}
{4.1743,1.4728, -3.7822}	{0.1872,0.1760,0.6860}	{-1.4187,1.3348, 6.2238}
{2.0076,3.9700, -3.6144}	{0.1765,0.1469,0.5325}	{1.5912,6.5068, -10.0293}
{1.8354,0.5612, -0.4329}	{0.9661,0.7067,0.4486}	{1.8354,7.2804, -0.4329}
{-1.2904,1.6430, -0.9800}	{0.7380,0.2766,0.6725}	{4.1821,4.6182, -2.9800}
{-3.4342,-2.2697, -3.3660}	{0.3779,0.5038,0.3419}	{3.8583, -2.2697,1.1621}
{-4.5445,2.4167, 3.6767}	{0.8852,0.2450,0.1191}	{-4.5445,2.4167, 3.6767}
{0.4069,-0.9304, -1.5523}	{0.2089,0.1257,0.1187}	{0.4069,-0.9304, -1.5523}
{1.2927, -3.3910,3.2671}	{0.2590,0.2771,0.6586}	{1.2927, -3.3910,3.2671}
{0.0910, -0.6836,3.3921}	{0.2989,0.2740,0.1336}	{0.0910, -0.6836,3.3921}

(For particle 1 and the third dimension:

(for third dimension)

```
trail_rand_num = 0.2019
CR = 0.5
```

As `trail _ rand _ num<` `CR` we have to apply **mutation:**

```
trail_num(particleindex,trail_selection_index) =
oldpop(num_c,trail_selection_index)+F*(oldpop(num_a,trail_se-
lection_index) - oldpop(num_b,trail_selection_index)) =
oldpop(7,3) + 1*(oldpop(8,3) - oldpop(9,3)) = 3.6767 +
(-1.5523 - 3.2671)
= -1.1426
```

So, trial vector for first individual will be

{1.9998, 0.8246, -1.1426}

Trail vectors are generated when the same process is repeated again and again for the whole solution set.

4.4.3 Selecting the Population for the Next Iteration

After defining the trail vector, the selection is between the parent vector and the trail vector to produce new offspring. During selection, the solution with the minimum

TABLE 4.5
New Population Generated after Applying Selection

Old Population,Xi	Function Cost (Xi)	Trial Vector X1	Function Cost (X1)	Final Population	Final Fitness
{1.9998,0.8246,4.3788}	112.497	{1.9998,0.8246,-1.1426}	86.6334	{1.9998,0.8246,-1.1426}	86.633
{4.1743,1.4728,-3.7822}	111.126	{-1.4187,1.3348,6.2238}	111.126	{4.1743,1.4728,-3.7822}	111.13
{2.0076,3.9700,-3.6144}	117.198	{1.5912,6.5068,-10.0293}	117.198	{2.0076,3.9700,-3.6144}	117.2
{1.8354,0.5612,-0.4329}	85.0209	{1.8354,7.2804,-0.4329}	85.0209	{1.8354,0.5612,-0.4329}	85.021
{-1.2904,1.6430,-2.9800}	94.7899	{4.1821,4.6182,-2.9800}	94.7899	{-1.2904,1.6430,-2.9800}	94.79
{-3.4342,-2.2697,-3.3660}	101.292	{3.8583,-2.2697,1.1621}	97.2758	{3.8583,-2.2697,1.1621}	97.276
{-4.5445,2.4167,3.6767}	134.631	{-4.5445,2.4167,3.6767}	102.653	{-4.5445,-0.7601,-0.7500}	102.65
{0.4069,-0.9304,-1.5523}	80.241	{0.4069,-0.9304,-1.5523}	80.241	{0.4069,-0.9304,-1.5523}	80.241
{1.2927,-3.3910,3.2671}	101.482	{1.2927,-3.3910,3.2671}	101.482	{1.2927,-3.3910,3.2671}	101.48
{0.0910,-0.6836,3.3921}	96.6725	{0.0910,-0.6836,3.3921}	96.6725	{0.0910,-0.6836,3.3921}	96.673

fitness is selected as offspring. See, for example, the first solution in which the parent vector is chosen as an offspring.

These selected solutions are then worked as the new offspring population. In the next iteration, these solutions will be converted into offspring.

5 Particle Swarm Optimization

5.1 INTRODUCTION

Many nature-inspired algorithms have been developed to address optimization problems. Some are inspired by natural selection and evolution theories, while others are motivated by swarm intelligence. Swarm intelligent algorithms are based on swarm intelligence. Particle swarm optimization (PSO), grey wolf optimization (GWO) algorithm, ant colony optimization (ACO), ant–bee colony, and cuckoo search are examples of algorithms in this area.

5.2 PSO ALGORITHMS

In comparison to other evolutionary algorithms, the PSO program is relatively simple. It was first proposed in 1995 by James Kennedy and Eberhart. The PSO algorithm was created by mapping the intelligence of swarms. PSO may appear to be a complicated technique, but it is actually quite simple. In PSOs, a set of randomly generated positions are changed over time to get a solution very close to optimum.

5.2.1 Natural Phenomena Used for Designing the Algorithm

PSO is a population-based algorithm that uses the process of bird flocking to solve optimization problems. To understand bird flocking, consider a group of birds moving in an area to find a hidden food source (see Figure 5.1). Each bird assesses the intensity of odor emanating from the food source to determine which direction they must move to capture the food source as soon as possible. The one who receives the most aroma chirps the loudest, attracting other birds to swing around in his direction. If any of the other circling birds get closer to the target than the first, it chirps louder, and the others veer toward it. The group of birds continues this behavior until one of the birds reaches the food source. The flocking process of birds represents intelligent behavior displayed by a group of birds while in search of food. Every bird updates its position and velocity while searching for food so that it can arrive at the food source as soon as possible. To update its position, each bird seeks advice from the bird with the most information about the food (referred to as `gBest` in the program) and examines its search history to determine the best position that it has ever attended during its search (referred to in the program as `pBest`). After comparing the positions of `gBest` and `pBest`, each bird calculates a new velocity and position.

DOI: 10.1201/9781003313649-5

FIGURE 5.1 An image showing the process of birds flocking.

5.2.2 Understanding the Relevance of Phenomena with Optimization

Now we'll look at how this method can help you solve an optimization problem. As discussed in the previous chapter, a randomized algorithm can solve an optimization problem if it can scan the entire search space and identify all potential areas that, when properly exploited, could produce optimal solutions. The scanning of the whole search space in a PSO is done with the assistance of birds. It's referred to as a solution in the algorithm. During scanning, suitable areas are identified by marking the values of `pbest`, and the best `pbest` solution is chosen as the global optimal solution. The operator in charge of scanning the entire search space is said to perform search space exploration. Scanning in a PSO is accomplished through the use of two operators known as the velocity and position operators. On the other hand, the exploration performed by these operators depends on random parameters used in defining equations. Identifying the correct values of these parameters that will result in a global optimal solution to a problem is a very common problem. Parameter

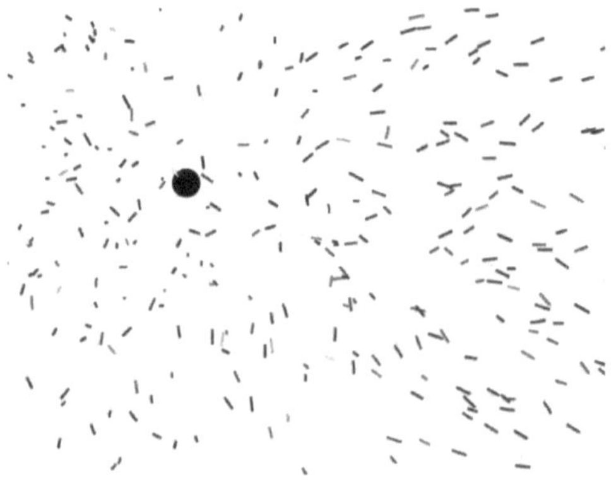

FIGURE 5.2A The view of an initialized population.

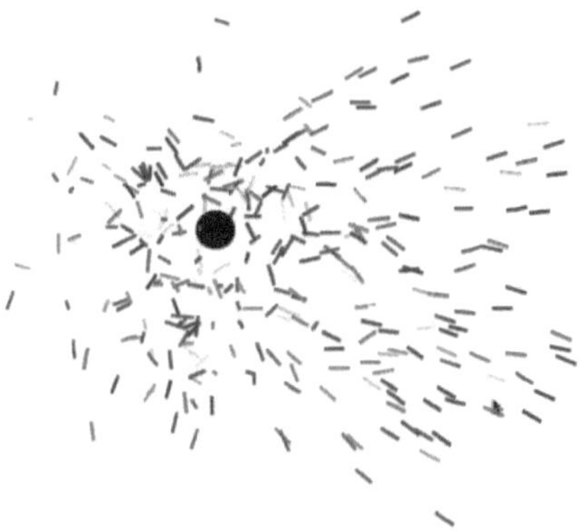

FIGURE 5.2B The view of an intermediate population. See how the swarm finds the food/prey.

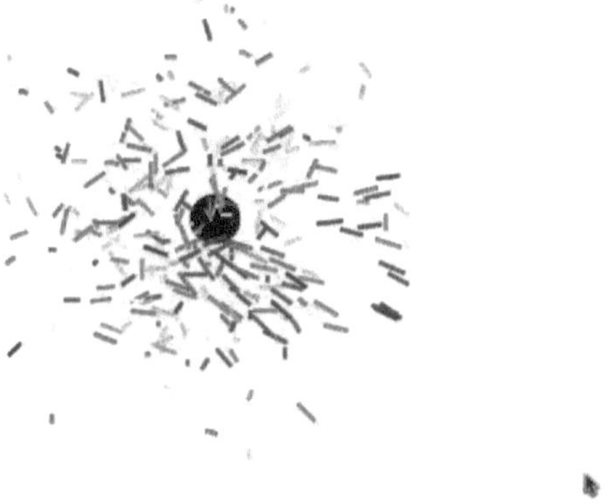

FIGURE 5.2C The view of an intermediate population. See how the swarm finds the food/prey.

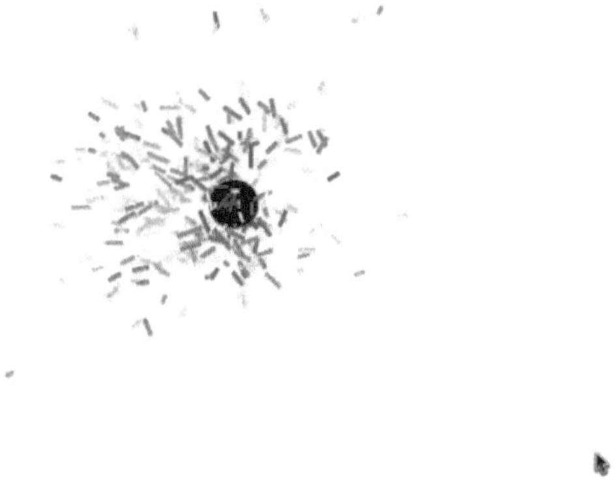

FIGURE 5.2D The view of an intermediate population. See how the swarm finds the food/prey.

FIGURE 5.2E The view of an intermediate population. See how the swarm finds the food/prey.

FIGURE 5.2F The view of an intermediate population. See how the swarm finds the food/prey.

FIGURE 5.2G The view of an intermediate population. See how the swarm finds the food/prey.

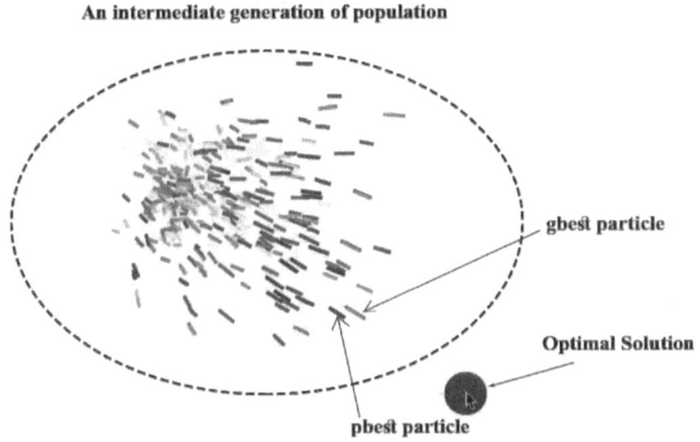

FIGURE 5.2H Swarm reaction to a changed objective; here you can clearly determine pbest and gbest.

tuning is another name for it. Because of incorrect values for these parameters, the algorithm is frequently stuck in a local optimal solution. To select `pbest` from the entire search space, use the selection operator. As a result, selection will act as an exploitation agent. A proper balance between exploitation and exploration is required to improve the performance of the algorithm.

5.3 DETAILED IMPLEMENTATION OF A PSO PROGRAM

To implement the process of bird flocking as a computer program, we must define the movement of birds during their food search in the form of functions/operators. Birds in the search space are represented in the program as particles or solutions. The movement of birds in the search space is defined in stages. Iterations refer to these steps. Particle velocities and positions are updated during each iteration to obtain new velocities and positions. The method for updating the particle position and the velocity in a given iteration is illustrated next.

5.3.1 Defining Velocities and Positions of Particles

To begin with, when no information about the particles is available, the initial position and velocity of the particles are generated at random from the given input domain. These solutions will represent the birds' initial positions and speeds when particle velocities and positions are available. They are supplied to obtain new particle velocities and positions. In the first iteration, for example, no information about the velocity and position of particles is available. As a result, these values are drawn at random from the given search domain. After completing the first iteration, the output of the first iteration will be supplied as an input to the second iteration. Similarly, the output of the second iteration will be supplied as an input to the third iteration.

5.3.2 Calculating gbest and pbest

After defining the velocities and positions of particles, the `gbest` and `pbest` are identified to generate new velocities and positions of particles. The fitness of each solution is calculated using the fitness function and are used to determine the `pbest` and `gbest` positions. Following the fitness calculation, each solution is guided with the help of global best and personal best. Because, during the flocking process, each bird tries to minimize the distance between itself and the food source. They follow the position of `gbest` and `pbest`. The minimum value is regarded as `gbest`. In each iteration, the position of the solution with the lowest fitness value is known as the global best. Similarly, each particle maintains its history of movement during different iterations; when defining a new position, they will check the best value of its history, which is referred to as personal best (`pbest`). Updating velocity and position of each particle produces the next generation.

Following the calculation of `pbest` and `gbest` positions, new particle positions and velocities are calculated. These values are computed using predefined equations that are used to update each particle's velocity and position. The PSO process can be defined mathematically using the following explanation. Assume that the search space is n-dimensional and that two vectors, namely, a position vector and a velocity vector, are required to implement the PSO program. An n-dimensional vector $X_i = \left(x_i^1, x_i^2, \ldots\ldots, x_i^n\right)$ can be used to represent the position of a swarm particle. Similarly, the velocity of the ith particle can be represented by another n-dimensional vector $V_i = \left(v_i^1, v_i^2, \ldots\ldots, v_i^n\right)$.

Additionally, the positions of global best solution and particle best solution are required to guide particle movement. When calculating `gbest` and `pbest`, the fitness of particles is considered. The value of its position vector determines each particle's fitness. Velocity vectors are never considered. They are simply used to define the particles' new positions. The particle's *i*th best position is saved in two vectors: personal best (`pbest`) and global best (`gbest`). `pbest` saves the best individual position $P_i = \left(p_i^1, p_i^2, \ldots\ldots, p_i^n\right)$ of each solution *i*th visited during its search. The position of the best individual of the whole swarm is noted as the global best (`gbest`) position $G = \left(g_1, g_2, \ldots\ldots, g_n\right)$. At each step, the velocity of the *i*th particle and its new position will be assigned according to the following two Equations 5.1 and 5.2:

$$V_i = \omega * V_i + c_1 * r_1 * \left(P_i - X_i\right) + c_2 * r_2 * \left(G - X_i\right) \tag{5.1}$$

$$X_i = X_i + V_i, \tag{5.2}$$

where ω is the inertia weight that controls the impact of a particle's previous velocity on its current velocity. r_1 and r_2 are randomly distributed random variables with independent uniform distributions and ranges (0,1). c_1 and c_2 are positive constant parameters known as acceleration coefficients that determine the maximum step size.

5.3.3 Clamping Velocity and the Position of Each Particle

In a PSO, the new velocity of a particle is calculated based on its previous velocity and the distance of its current position from both its own best historical position and the best position of the entire population or its neighborhood. Generally, the value of each component in V can be clamped to the range `[-vmax, vmax]` to control the excessive roaming of particles outside the search space. Then the particle flies toward a new position according to Equation 5.2. Clamping of the position vector is also done to make the values of a variable in given bounds.

5.3.4 Termination PSO Program

There are two ways to stop the algorithm. In the first method, the algorithm is stopped once the desired value is obtained in any iteration. In other forms, the position of particles is modified until the count reaches the maximum allowable iterations for the algorithm.

5.3.5 Parameter Tuning

Some parameters, such as w, c_1, c_2, are essential in defining the convergence of the PSO algorithm. By varying the values of these parameters, the performance of the PSO algorithm can be altered. Parameter tuning is the process of altering the performance of PSO by changing the values of these parameters. Setting these parameters is dependent on the expert's knowledge of the problem. Setting these parameters in some PSO programs does not require the assistance of an expert; instead, the program does it automatically. Such algorithms are known as self-adaptive algorithms.

5.3.6 Pseudocode of a PSO Algorithm

The pseudocode for a PSO is represented as follows; the pseudocodes used in the algorithm's design listed in Table 5.1.

TABLE 5.1
Abbreviations Used in Pseudocodes

Notations	Description
`x`	Current population
`v`	Current velocity
`c1, c2`	Constants
`w`	Inertia weight
`Mgen`	Maximum number of generations
`pbest`	Personal best position for each particle
`gbest`	Generation best position
`ps`	Population size
`Dim`	Dimension

Algorithm 5.1: Algorithm of a PSO

```
PSO()
begin
x = rand(ps,Dim) // Initialize the particle's position with a random vector
v = rand(ps,Dim) // Initialize the velocity of the particles
pbest = X //Initialize the particle's best-known position to its initial position
/*evaluate the function cost for each particle*/
gbest = particle having minimum cost // update the swarm's best-known position
do
For each i = 1 to Mgen do
v = w*v+c1*rand()*(pbest-x)+c2*rand()*(gbest-x)// Update the particle's velocity
x = x+v // Update the particle's position
cost _ x = evaluate the function cost for new particles
for each individual in pbest and x do
If (cost(Xi) < cost(pbesti)) // Update the particle's best known position
Pbesti = xi
endif
if(cost(pbesti) < cost(gbest)) // Update the swarm's best-known position
gbest = pbesti
endif
end
end
until stopping criterion is met
End
```

5.4 MATLAB® IMPLEMENTATION OF A PSO PROGRAM

Let us now look at the MATLAB implementation of the PSO program. Each MATLAB instruction's purpose is explained in detail. After reading this section, you will be able to write other similar programs using the instructions used to implement PSO.

Defining initial population:

The first step in writing any randomized algorithm is to define the program's initial population. Some solutions generated from the entire search space are included in the initial population. The initial population size in complex problems is 40. However, to discuss the PSO algorithm in this program, we have used ten solutions in the initial

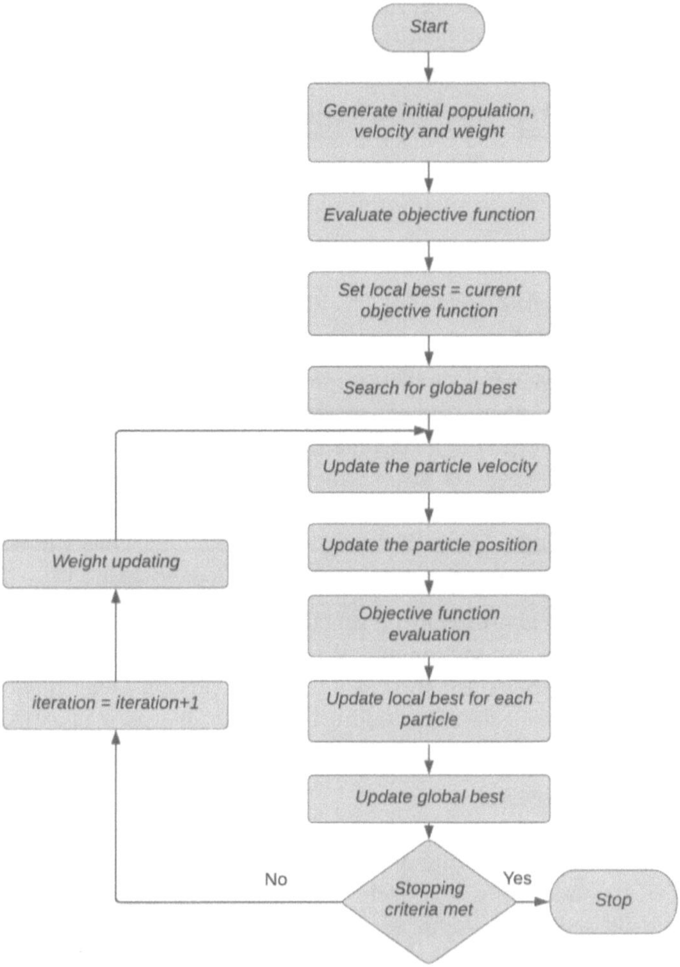

FIGURE 5.3 Flow Chart of a PSO Algorithm

population. In MATLAB, the population can be started by declaring a variable and then assigning it a value. The variable name `popsize` is used in this program to define the population size, and it can be initialized simply by writing this command.

```
popsize = 10
```

Dimension and fitness function:

After populating the population, we must work according to the algorithm's theory. In a PSO, solutions are represented by position values, which are updated based on the new velocity of these particles. As a result, in order to define solutions and update their positions, we must first determine the initial position and velocity of each particle/solution. In the initial population of PSO, both position and velocity will be randomly initialized. Because PSO is used to solve an optimization problem, we want to identify the value of the optimal solution that will produce the optimal value of the function. We have to check how many variables will be used for defining the solution vector. These variables define the dimension of the solution vector.

Let us take an example and check the dimension of the solution vector.

The function that we want to optimize is known as the fitness function. Let us suppose that we want to optimize (minimize) a function *F*, which is defined as in Equation 5.3:

$$Minimize\, F\left(x_1,\, x_2,\, x_3\right) = x_1^2 + x_2^2 + x_3^2. \tag{5.3}$$

It is a multivariable function in which three variables represent each solution. This problem's fitness function is a three-variable function whose value will be determined by calculating $F(x_1, x_2, x_3)$. Assume that the optimal solution of this function is `xopt` and that three variables also represent it. The dimension of the solution for this problem will be 3. Variables in optimization problems are bounded, with both a lower and an upper bound. In the current problem, each variable's lower and upper bounds are –5 and 5, respectively. In other cases, it may be different.

Checkpoint 1: Write a two-dimension function.
Answer: A two-dimension function has two input variables. Therefore, the area of a rectangle is a two-dimension function as it uses two variables length and width.

`A(l,b) =l*b` where l is length and b is breadth; hence, it is a two-dimension function.

Defining boundary vectors:
In our example, we have a three-variable function with lower and upper bounds of –5 and 5, respectively. A minimum vector, also known as an `xmin`, is commonly used to define a solution in which each variable is represented by the lower bound. Similarly, `xmax` is the maximum vector in which all variables are at their upper bounds. In MATLAB, we can use the following steps for defining minimum and maximum vectors since each variable is at the lower bound in the minimum vector, which in this example is –5.

We can define two variables `xbound` and `vbound`, these vectors are used to create boundary vectors for position and velocity.

After defining these variables, we can use a MATLAB command `ones(row, column)` to generate a matrix of row `x` column `v` in which the value of all variables will be 1. Row variable specifies the number of rows and column variable `v` specifies the number of columns.

For example, let us suppose that we want to generate a three-variable vector in which all variables are equal, then this can be done by writing the following command:

```
ones(1,3) = {1, 1, 1}
```

After generating one vector, we can multiply it with `xbound = 5` and `vbound = 5` to create maximum and minimum vectors.

For example:

```
Xmin=- xbound*ones(1,DIM)
Xmax=xbound*ones(1,DIM)
```

Similarly

```
Vmin=- vbound*ones(1,DIM)
Vmax=vbound*ones(1,DIM)
```

In some problems, we can use separate bounds for position and velocities.

5.4.1 Population Initialization

Once the maximum and minimum vectors are generated. We can develop the initial population. Because the initial population size is ten, we must generate ten position and velocity vectors. These vectors will be generated within the given variable boundaries. The initial population's position and velocity vector will define the input to the algorithm's first iteration.

In MATLAB, we can generate random values for all solutions in the form of a matrix with a single statement `rand(row, column)`. However, in our problem, these random values will be generated between (0 and 1), where `0` is the lower bound and `1` is the upper bound.

We need to generate interval values `(-5,5)`. This can be generated as follows.

First iteration:
First, we define a `t` vector and use following statement:

```
t1 = rand(popsize,DIM)
```

Assume we want to define the values of ten solution vectors, with three variables representing each solution. Then we must set `popsize` to `10` and `DIM` to `3`, and use the preceding command.

Although this will generate 10 solutions of three dimensions, the values of each dimension will be between `(0,1)`. However, as per our requirement, we need to

TABLE 5.2
Random Values between 0 and 1

{0.9844, 0.4479, 0.5368}
{0.2087, 0.3773, 0.7255}
{0.6239, 0.5901, 0.3954}
{0.4681, 0.0423, 0.9275}
{0.6825, 0.1180, 0.9136}
{0.8447, 0.7823, 0.2330}
{0.0348, 0.1755, 0.0892}
{0.3474, 0.8594, 0.8258}
{0.0872, 0.6357, 0.7281}
{0.4911, 0.7154, 0.1372}

TABLE 5.3
Initial Positions

{4.8439,	-0.5211,	0.3675}
{-2.9132,	-1.2274,	2.2552}
{1.2393,	0.9007,	-1.0457}
{-0.3185,	-4.5774,	4.2748}
{1.8252,	-3.8203,	4.1358}
{3.4469,	2.8225,	-2.6702}
{-4.6520,	-3.2447,	-4.1077}
{-1.5264,	3.5941,	3.2585}
{-4.1276,	1.3566,	2.2812}
{-0.0894,	2.1538,	-3.6282}

generate random values between `-5 to 5`. For developing random values in given bounds, we have to multiply our `t` matrix by `10`. This will create a matrix in which all values will be between `0` and `10`. Now for generating it between `-5 to 5`, we will subtract `5` from each matrix value.

So this equation can be used to generate position vectors.

$$x = 2 * xbound * t1 - xbound \tag{5.4}$$

What is accomplished with the assistance of this function? To begin, we multiply each matrix value by `10`. As a result, the generated random values will be domain-specific `(0,10)`. We will subtract `xbound` from each value to keep it between `(-5 and 5)`. We can develop the initial value of the position between `(-5,5)` in this manner.

Similarly, we will use similar equations to generate velocities for all particles.

First, a vector `t2` is generated by a similar command.

```
t2 = rand(popsize,DIM)
```

TABLE 5.4
Random Values Generated for Velocity

{0.2861,	0.9343,	0.5498}
{0.4218,	0.1116,	0.1330}
{0.7628,	0.7652,	0.9445}
{0.7519,	0.1680,	0.8199}
{0.1220,	0.1158,	0.5776}
{0.8721,	0.7850,	0.2261}
{0.7558,	0.2196,	0.8201}
{0.7446,	0.0663,	0.2961}
{0.5473,	0.5192,	0.6502}
{0.1698,	0.4197,	0.2356}

TABLE 5.5
Initial Positions and Velocities of Particles

Particles	Velocity
{4.8439, -0.5211, 0.3675}	{-2.1391, 4.3433, 0.4979}
{-2.9132, -1.2274, 2.2552}	{-0.7819, -3.8836, -3.6702}
{1.2393, 0.9007, -1.0457}	{2.6277, 2.6524, 4.4446}
{-0.3185, -4.5774, 4.2748}	{2.5193, -3.3196, 3.1989}
{1.8252, -3.8203, 4.1358}	{-3.7796, -3.8419, 0.7761}
{3.4469, 2.8225, -2.6702}	{3.7211, 2.8499, -2.7388}
{-4.6520, -3.2447, -4.1077}	{2.5582, -2.8042, 3.2013}
{-1.5264, 3.5941, 3.2585}	{2.4456, -4.3373, -2.0387}
{-4.1276, 1.3566, 2.2812}	{0.4726, 0.1919, 1.5019}
{-0.0894, 2.1538, -3.6282}	{-3.3023, -0.8026, -2.6438}

Then for keeping the values in (–5, 5), the following command is used:

```
v = 2 * vbound * t2 -vbound
```

The initial position and velocity vector are represented in Table 5.5.

Checkpoint 2: Write a MATLAB code to generate a population of `20` solutions, each with three variables.

Answer:

Let us suppose that the lower bound and upper bound vectors are defined as follows:

`UL = [5.1200 5.1200 5.1200]` = Upper Bound
`LL = [-5.1200 -5.1200 -5.1200]` = Lower Bound]

To generate a random matrix of size `(20 x 3)`, we have to write:

```
t1=rand(20,3)
```

To generate solutions in the given bound a new vector `t2` should be formed by using following equation:

```
t2=LL+(UL-LL).*t1;
```

This will create a matrix of size `20 x 3` in which each variable will be between `-5.12 to 5.12`.

5.4.2 Calculating gbest and pbest

To generate new position and velocity vector values, we must first determine the `pbest` and `gbest` values from the input. `gbest` is the best solution's solution vector. `pbest` value is yet another value that is used in the creation of new positions. Each solution maintains `pbest` values and stores the solution vector of the best

TABLE 5.6
Fitness of Initial Particles with gbest Position

Particles	Function Cost	gbest
{4.8439, -0.5211, 0.3675}	102.1532	
{-2.9132, -1.2274, 2.2552}	98.3845	
{1.2393, 0.9007, -1.0457}	**84.7899**	
{-0.3185, -4.5774, 4.2748}	116.4954	
{1.8252, -3.8203, 4.1358}	112.6638	**{1.2393, 0.9007, -1.0457}**
{3.4469, 2.8225, -2.6702}	109.3049	
{-4.6520, -3.2447, -4.1077}	119.3456	
{-1.5264, 3.5941, 3.2585}	121.0771	
{-4.1276, 1.3566, 2.2812}	114.0155	
{-0.0894, 2.1538, -3.6282}	98.9918	

position of a particle attained during its search in previous generations. To determine the `gbest` and `pbest` values, we must first evaluate the position of each vector. Look, we can't measure velocity. Because this is the first iteration, there will be no movement history. So only initial positions will be counted to calculate the values of `pbest`. In the initial generation, `pbest` will be equal to `x`, and the position of the `gbest` solution can be calculated from the `pbest` solution. The solution with a minimum value of fitness function will define `gbest`.

The position of `gbest` is highlighted with bold values in Table 5.6.

```
pbest = x =
```

5.4.3 Updating Velocity and Position Vectors

Second iteration:

In the second iteration, we have to calculate new velocities for all the particles, and then these updated velocities will be used to define new positions for all particles.

In a PSO, the following expression is used to update the velocity vector of each particle:

$$v = w * v + c1 * rand_1 * (pbest - x) + c2 * rand_2 * (gbest - x). \tag{5.5}$$

In MATLAB, we can update the velocities of all particles in a single statement. However, for updating all solutions in a single go, the variables used in the formula should be of the same size. We need to use the same size of matrices for all variables.

For example, if we want to update the velocities of all particles at once, we must first determine how many new velocity vectors will be generated. The population size and the dimension of the velocity matrix will select the size of the matrix. The size of the velocity matrix, as well as the position matrix, will be `(popsize,DIM)`.

We generated two vectors, v, and x, in the previous generation, and the size of these vectors was also (popsize,DIM).

As a result, we must use the same size matrix to update all particle velocities in a single equation. We can see from the previous values that the sizes of pbest and x are the same, so they can be easily subtracted.

The size of gbest, on the other hand, is (1, Dim), and the size of x is (popsize,DIM).

The repmat function can be used to create gbests of the same size.

repmat(x,row,column) where x is a matrix, row defines the size of the row and column determines the size of the column.

repmat is used to iteratively repeat the value of a matrix for the specified number of times.

Because we need to generate ten different values of gbest to update the particle velocities.

```
repmat(gbest,popize,1)=repmat will be written (gbest,popize,1)
```

This will generate popsize copies of gbest.

t3 and t4 are random vectors of the same size that can be generated using the same equations. The program's output values are shown in Table 5.7. Because the values of random vectors must be multiplied by other matrix elements one by one, the term product is used in the equation. In this case, matrix multiplication is unnecessary.

```
t3 = rand(popsize,DIM) =
t4 = rand(popsize,DIM) =
t3= repmat(gbest,popsize,1) =
```

New velocities after calculation are shown in Table 5.10. They are calculated by using the equation v = w*v + c1*t1.*(pbest-x) + c2*t2.*(t3-x).

TABLE 5.7
Random Values Generated for r1

{0.2779, 0.6988, 0.6648}
{0.2636, 0.0861, 0.0692}
{0.2388, 0.5977, 0.4304}
{0.6980, 0.1227, 0.1427}
{0.0586, 0.2303, 0.4992}
{0.4566, 0.5666, 0.4611}
{0.2843, 0.7702, 0.9858}
{0.1536, 0.4907, 0.4687}
{0.7446, 0.1801, 0.2824}
{0.9511, 0.3382, 0.2353}

TABLE 5.8
Random Values Generated for r2

{0.5202, 0.4773, 0.5669}
{0.4274, 0.0442, 0.0291}
{0.9435, 0.7869, 0.0135}
{0.8878, 0.8224, 0.0510}
{0.1261, 0.2335, 0.6905}
{0.4247, 0.4768, 0.1267}
{0.8188, 0.9047, 0.9138}
{0.9066, 0.4857, 0.8297}
{0.9064, 0.6807, 0.3047}
{0.8325, 0.2990, 0.5245}

TABLE 5.9
Copies of gbest Solutions

{1.2393, 0.9007, -1.0457}
{1.2393, 0.9007, -1.0457}
{1.2393, 0.9007, -1.0457}
{1.2393, 0.9007, -1.0457}
{1.2393, 0.9007, -1.0457}
{1.2393, 0.9007, -1.0457}
{1.2393, 0.9007, -1.0457}
{1.2393, 0.9007, -1.0457}
{1.2393, 0.9007, -1.0457}
{1.2393, 0.9007, -1.0457}

TABLE 5.10
Updated Velocity of Particles

Velocity
{-4.4965, 4.4541, -0.8029}
{2.0327, -2.9354, -3.0502}
{2.0811, 2.1007, 3.5201}
{4.0622, 4.1034, 2.1284}
{-3.1039, -1.3951, -4.7319}
{1.5461, 0.8877, -1.8615}
{9.2350, 3.3837, 6.7169}
{5.6838, -5.3900, -6.9512}
{7.6441, -0.3117, -0.3254}
{-0.9625, -1.1956, -0.0697}

5.4.4 Clamping Velocity and Position Vectors

Clamping techniques are used to convert all variable values within the defined lower and higher boundaries after velocities have been calculated. Because the lower and upper bounds of each variable are `(-5,5)`, we start by counting how many variables are out of bounds. We will use two equations to figure out how many values are fewer than the lowest velocity value and how many are greater. Other can also be used to find out how many values are greater than the maximum. After they have been checked, their boundary values will take the place of these values. First, I explain how velocity values less than the minimum are identified and changed. A similar approach will be used for the identification of velocity values that are greater than the maximum and replaced.

For the identification of values that are less than the minimum, we will be using equation `s = v <repmat(vmin,popsize,1)`.

By this equation, we compare the new velocity vector with the min value of the velocity. This equation will generate a Boolean matrix `s`, where `0` will be stored when the value is in bounds and `1` will be stored when this condition becomes false, or the value is out of bounds.

The value of `s` for given an example will be as per Table 5.11.

After checking which values are less than the minimum, the following equation will be used to replace that by the minimum value of the velocity. Look only at the s values; those that are 1 will be replaced by the minimum value of the velocity.

$$v = (1 - s) * v + s * repmat(vmin, popsize, 1) \tag{5.6}$$

Once all velocity values that are less than the minimum are replaced, we will use the same method for replacing values that are greater than the upper value.

Like the `s` matrix, we calculate `b` matrix in which `0` will be placed when the value is in bounds or less than the upper bound and `1` when the value is greater than the upper bound.

TABLE 5.11
Checking out of bound values

{0, 0, 0}
{0, 0, 0}
{0, 0, 0}
{0, 0, 0}
{0, 0, 0}
{0, 0, 0}
{0, 0, 0}
{0, 1, 1}
{0, 0, 0}
{0, 0, 0}

TABLE 5.12
Updated Velocity of Particles

{-4.4965, 4.4541, -0.8029}
{2.0327, -2.9354, -3.0502}
{2.0811, 2.1007, 3.5201}
{4.0622, 4.1034, 2.1284}
{-3.1039, -1.3951, -4.7319}
{1.5461, 0.8877, -1.8615}
{9.2350, 3.3837, 6.7169}
{5.6838, **-5.0000,-5.0000**}
{7.6441, -0.3117, -0.3254}
{-0.9625, -1.1956, -0.0697}

Note: Values in bold represents values obtained after clamping.

TABLE 5.13
Checking Velocity Vector Greater than vmax

{0, 0, 0}
{0, 0, 0}
{0, 0, 0}
{0, 0, 0}
{0, 0, 0}
{0, 0, 0}
{1, 0, 1}
{1, 0, 0}
{1, 0, 0}
{0, 0, 0}

For our example, the value of b vector will be

```
b = v > repmat(vmax,popsize,1)
```

Once b matrix is calculated, we will replace these values by maximum value of the velocity.

```
v = (1-b).*v + b.*repmat(vmax,popsize,1)
```

After velocity clamping, we will add the new velocity in the previous position and generate a new position x. Updated position for our example is shown in Table 5.15.

Updated Position

```
x = x+v;
```

After calculating new positions, the clamping of each position will be done to make it in bound.

TABLE 5.14
Updating Velocity of Particles

{-4.4965, 4.4541, -0.8029}
{2.0327, -2.9354, -3.0502}
{2.0811, 2.1007, 3.5201}
{4.0622, 4.1034, 2.1284}
{-3.1039, -1.3951, -4.7319}
{1.5461, 0.8877, -1.8615}
{5.0000, 3.3837, 5.0000}
{5.0000, -5.0000, -5.0000}
{5.0000, -0.3117, -0.3254}
{-0.9625, -1.1956, -0.0697}

TABLE 5.15
Updated Positions of Particles

Old Particle Position	Velocity	New Particle Position
{4.8439, -0.5211,0.3675}	{-4.4965,4.4541, -0.8029}	{0.3474,3.9330, -0.4354}
{-2.9132,-1.2274,2.2552}	{2.0327,-2.9354, -3.0502}	{-0.8805, -4.1627,-0.7950}
{1.2393,0.9007, -1.0457}	{2.0811,2.1007, 3.5201}	{3.3204,3.0014, 2.4744}
{-0.3185,-4.5774, 4.2748}	{4.0622,4.1034, 2.1284}	{3.7436,-0.4740, 6.4032}
{1.8252,-3.8203, 4.1358}	{-3.1039,-1.3951, -4.7319}	{-1.2787, -5.2154,-0.5961}
{3.4469, 2.8225, -2.6702}	{1.5461, 0.8877, -1.8615}	{4.9929, 3.7102, -4.5318}
{-4.6820,-3.2447, -4.1077}	{5.0000, 3.3837, 5.0000}	{0.3180, 0.1390, 0.8923}
{-1.5264, 3.5941, 3.2585}	{5.0000,-5.0000, -5.0000}	{3.4736,-1.4059, -1.7415}
{-4.1276, 1.3566, 2.2812}	{5.0000,-0.3117, -0.3254}	{0.8724, 1.0448, 1.9558}
{-0.0894, 2.1538,-3.6282}	{-0.9625,-1.1956, -0.0697}	{-1.0519, 0.9583,-3.6979}

We will apply the same procedure that we used for velocity.

Clamping Particle Position

```
s = x < repmat(xmin,popsize,1)
```

TABLE 5.16
Checking Out-of-Bound Positions

```
{0, 0, 0}
{0, 0, 0}
{0, 0, 0}
{0, 0, 0}
{0, 1, 0}
{0, 0, 0}
{0, 0, 0}
{0, 0, 0}
{0, 0, 0}
{0, 0, 0}
```

TABLE 5.17
Updating Out-of-Bound Positions

{0.3474,	3.9330,	-0.4354}
{-0.8805,	-4.1627,	-0.7950}
{3.3204,	3.0014,	2.4744}
{3.7436,	-0.4740,	6.4032}
{-1.2787,	**-5.0000,**	-0.5961}
{4.9929,	3.7102,	-4.5318}
{0.3480,	0.1390,	0.8923}
{3.4736,	-1.4059,	-1.7415}
{0.8724,	1.0448,	1.9558}
{-1.0519,	0.9583,	-3.6979}

Note: The values in bold are clamped values.

The new position vector after clamping values that are less than minimum.

```
x = (1-s).*x + s.*repmat(xmin,popsize,1)
b = x > repmat(xmax,popsize,1)
```

New position vector after clamping values greater than minimum.

```
x = (1-b).*x + b.*repmat(xmax,popsize,1)
b = Boolean variable for clamping
```

After calculating new position values, the objective function values at each position are calculated.

```
cost_x =
```

TABLE 5.18
Checking Velocity Values That Are Greater Than the Max Value

{0,	0,	0}
{0,	0,	0}
{0,	0,	0}
{0,	0,	1}
{0,	0,	0}
{0,	0,	0}
{0,	0,	0}
{0,	0,	0}
{0,	0,	0}
{0,	0,	0}

TABLE 5.19
Updating Positions

{0.3474,	3.9330,	-0.4354}
{-0.8805,	-4.1627,	-0.7950}
{3.3204,	3.0014,	2.4744}
{3.7436,	-0.4740,	**5.0000**}
{-1.2787,	-5.0000,	-0.5961}
{4.9929,	3.7102,	-4.5318}
{0.3480,	0.1390,	0.8923}
{3.4736,	-1.4059,	-1.7415}
{0.8724,	1.0448,	1.9558}
{-1.0519,	0.9583,	-3.6979}

Note: The bold value depicts clamped values.

TABLE 5.20
Boolean Variable for Clamping

{0,	0,	0}
{0,	0,	0}
{0,	0,	0}
{0,	0,	1}
{0,	1,	0}
{0,	0,	0}
{0,	0,	0}
{0,	0,	0}
{0,	0,	0}
{0,	0,	0}

TABLE 5.21
Fitness Calculation of New Positions

New Particle Position	New Function Cost
{0.3474, 3.9330,-0.4354}	105.4783
{-0.8805,-4.1627,-0.7950}	89.8050
{3.3204, 3.0014, 2.4744}	116.4107
{3.7436,-0.4740, **5.0000**}	124.8962
{-1.2787,-5.0000,-0.5961}	96.6120
{4.9929, 3.7102,-4.5318}	140.1360
{0.3480, 0.1390, 0.8923}	83.7807
{3.4736,-1.4059,-1.7415}	90.9509
{0.8724, 1.0448, 1.9558}	91.8923
{-1.0519, 0.9583,-3.6979}	94.4998

Note: The bold value depicts clamped values.

Checkpoint 3: Assume we have the following function:

$$f(x_1, x_2) = 5 * x_1^2 - 7 * x_2^2 + 9,$$

where $2 <= x_1 <= 5$ and $1 <= x_2 <= 4$.

We want to apply a PSO to minimize $f(x_1, x_2)$. This is the first iteration. Check the initial set of solutions (say, in set 1) in the following table and answer the following questions; explain all necessary details.

Given the following four solutions

Solution	Values
P_1	(2,1)
P_2	(2,2)
P_3	(3,3)
P_4	(4,3)

which solution will represent the position of `gbest`?

a. Suppose new solutions (positions) that are generated after the first generation (assume that we have calculated new velocities and added to the previous positions), create a new set of four solutions that are simply the mirror image of initial values. What will be the new position of `gbest`? (**Note:** when we take mirror image of P_1, a new solution, P_5, with position (1,2) will be created, remember the values will be clamped when lies outside the input boundaries.)
b. Identify the values of `pbest`.

Answer:

a. The fitness value of each solution will be

Solution	Values(x_1,x_2)	Fitness $5*x_1^2 - 7 * x_2^2 + 9$	Gbest
P_1	(2,1)	5*4-7*1+9=22	
P_2	(2,2)	5*4-7*4+9=1	
P_3	**(3,3)**	**5*9-7*9+9=--9**	**(3,3)**
P_4	(4,3)	5*16-7*9+9=26	

b. New positions

Updating values of gbest and pbest for creating solution of a third generation: The new value of pbest is calculated by comparing the objective value of previous value of pbest and the objective value of new position. The new value of pbest will store the best value.

After calculating new values of pbest, the best of pbest will be chosen as gbest.

Calculating pbest:
The new value of pbest and gbest will be used to calculate the new velocity vector for the third generation.

This PSO process is repeated until either the best solution is obtained or the maximum generations are reached.

TABLE 5.22
Updated Positions after Clamping

Previous Positions	Values	New Positions (mirror image)	Mirror Values	After Clamping
P_1	(2,1)	P_5	(1,2)	(2,2)
P_2	(2,2)	P_6	(2,2)	(2,2)
P_3	(3,3)	P_7	(3,3)	(3,3)
P_4	(4,3)	P_8	(3,4)	(3,4)

TABLE 5.23
Updated Values of pbest

Previous Positions xp	Values	New Positions (mirror image)	New Positions xnp	Fitness Comparison Min (F(xp), F(xnp)	pbest
P_1	(2,1)	P_5	(2,2)	Min(22,1)=1	(2,2)
P_2	(2,2)	P_6	(2,2)	Min(1,1)=1	(2,2)
P_3	(3,3)	P_7	(3,3)	Min(-9,-9)=-9	(3,3)
P_4	(4,3)	P_8	(3,4)	Min(26,-48)=-48	(3,4)

TABLE 5.24
Upadated gbest

Old pbest	Function Cost for Old pbest Values	New Particle Positions	Function Cost for New Particles	pbest	Function Cost of pbest	gbest
{4.8439, -0.5211, 0.3675}	102.1532	{0.3474, 3.9330, -0.4354}	105.4783	{4.8439, -0.5211, 0.3675}	102.1532	
{-2.9132, -1.2274, 2.2552}	98.3845	{-0.8805, -4.1627, -0.7950}	89.8050	{-0.8805, -4.1627, -0.7950}	89.8050	
{1.2393, 0.9007, -1.0457}	84.7899	{3.3204, 3.0014, 2.4744}	116.4107	{1.2393, 0.9007, -1.0457}	84.7899	
{-0.3185, -4.5774, 4.2748}	116.4954	{3.7436, -0.4740, **5.0000**}	124.8962	{-0.3185, -4.5774, 4.2748}	116.4954	
{1.8252, -3.8203, 4.1358}	112.6638	{-1.2787, -5.0000, -0.5961}	96.6120	{-1.2787, -5.0000, -0.5961}	96.6120	{0.3480, 0.1390, 0.8923}
{3.4469, 2.8225, -2.6702}	109.3049	{4.9929, 3.7102, -4.5318}	140.1360	{3.4469, 2.8225, -2.6702}	109.3049	
{-4.6520, -3.2447, -4.1077}	119.3456	{0.3480, 0.1390, 0.8923}	83.7807	{0.3480, 0.1390, 0.8923}	83.7807	
{-1.5264, 3.5941, 3.2585}	121.0771	{3.4736, -1.4059, -1.7415}	90.9509	{3.4736, -1.4059, -1.7415}	90.9509	
{-4.1276, 1.3566, 2.2812}	114.0155	{0.8724, 1.0448, 1.9558}	91.8923	{0.8724, 1.0448, 1.9558}	91.8923	
{-0.0894, 2.1538, -3.6282}	98.9918	{-1.0519, 0.9583, -3.6979}	94.4998	{-1.0519, 0.9583, -3.6979}	94.4998	

Note: The bold value depicts clamped values.

5.5 IMPROVING THE PERFORMANCE OF A PSO ALGORITHM

We are always interested in improving the performance of existing algorithms. Although the theory of PSO is well defined and shows good performance on optimization problems, we can do some changes in PSO to create faster variants of the PSO algorithm. Two types of changes can improve the performance of the existing version. In the first method, we will not change the equations used in the different operators. The only thing that we will be doing is parameter tuning. In other methods, we can change the equations used in different operators and make it more suitable for solving problems.

5.6 REAL-LIFE APPLICATIONS OF A PSO ALGORITHM

A PSO algorithm can be used to solve many real-life problems. Some problems that belong to computer science are workflow scheduling problems in the cloud, test-case generation, and classification problems. We can apply PSO for economic load dispatch problems, dynamic load dispatch problems, and electrical smart-grid designing.

6 Grey Wolf Optimization

6.1 INTRODUCTION

Many swarm intelligence–based algorithms have been developed in the past to solve complex optimization problems. These swarm intelligence-based algorithms are classified into two groups. First-class algorithms, such as ant colony optimization (ACO) and ant–bee colony (ABC), use the intelligence of insects to solve optimization problems. Other categories include algorithms that mimic the intelligence of animals and birds, such as partial swarm optimization (PSO), grey wolf optimization (GWO), and many others.

6.2 GWO ALGORITHMS

GWO algorithm was proposed by Saiyadali Mirjalili in 2005 for solving optimization problems. Following in the footsteps of PSO, the GWO is gaining popularity as a swarm intelligence algorithm that solves optimization problems by mimicking the hunting process used by grey wolves. This algorithm is designed by simulating the hunting process used by a group of wolves to attack prey. Unlike other animals, grey wolves prefer to hunt in a group, and it's a guided activity in which wolves with different intelligence levels participate. Depending on their role in the hunting process, these wolves are categorized into four classes, that is, alpha, beta, delta, and omega, for creating the algorithm. Alpha, beta, and delta wolves use their experience to guide omega wolves in performing hunting. The hunting process used by a group of grey wolves is mapped in three steps. These steps are searching the prey, encircling the target, and attacking on the prey.

6.2.1 Natural Phenomena Used for Designing the Algorithm

As discussed in the previous paragraph, GWO maps the hunting process used by a group of grey wolves. To understand the hunting process used by grey wolves, we need to know about them and their characteristics and then discuss how they perform hunting. Grey wolves are on the top level of the food chain and are part of the top predators. They prefer to hunt in a group. The group size varies from 5 to 12. Based on work performed by grey wolves, they can be classified into four categories. The first category of wolves is known as alpha wolves. They work as leaders of the group and manage essential activities like hunting, sleeping, or walking. They are decision-makers and lead the entire group. The second category is beta wolves, who work under the alpha and help alpha wolves finalize decisions. They are also responsible for propagating the alpha's commands throughout the group and providing feedback to the alpha. The third category of wolves is delta, which works for the group. Unlike alpha and beta wolves, they perform much work. Many of them are scouts, sentinels, elders, hunters,

DOI: 10.1201/9781003313649-6

and caretakers. Finally, the lowest ranking wolf is called omega. These wolves assist all other dominant wolves, and they need to work according to the direction given by dominating members. For simply explaining the hunting process, the whole process is divided into three separate processes, that is, searching for prey, encircling prey, and attacking targets performed by wolves of different classes. During the search for prey, alpha, beta, and delta wolves identify the position of the prey. During enriching, omega wolves change their position according to the direction provided by alpha, beta, and delta wolves. After enriching, omega wolves attack prey.

6.2.2 Understanding the Relevance of Phenomena with Optimization

Now, we discuss how the hunting process used by grey wolves can help solve an optimization problem and how the hunting process can be converted into a randomized algorithm. As discussed in the previous chapter, for solving an optimization problem, a randomized algorithm must be able to scan the whole search space properly and exploit all probable areas where the optimal solution may lie. In GWO, the position of wolves will define whether exploitation is to be done or exploration will be implemented. For example, in the first generation of a GWO program, the initial position of wolves is generated randomly within the given lower and upper bounds of the variable. This initialization method will create some random solutions in the search space and help explore the search space. After creating solutions, their fitness value is evaluated, and the three best solutions are identified for exploitation and exploration. These three best solutions are then exploited and explored to investigate other good solutions in new iterations. The positions of wolves are updated in a step-by-step manner. Each step defines one iteration, which shows how the wolves are changing their positions.

In GWO, the position of the three best solutions is used to guide the new positions of the whole pack. The search space is scanned in each iteration, and the three best

FIGURE 6.1 A pack of wolves hunting down a snow reindeer.

places, that is, alpha, beta, and delta, are identified. These best positions will be stored as the position of the three best wolves and will be used to decide the new position of each wolf for the next iteration. These new positions of wolves will define the next generation of the algorithm. Again, the same process is used, and new alpha, beta, and delta values are identified by checking the fitness of new solutions. The operators responsible for generating new positions in the whole search space are said to perform exploration of the search space. In GWO, the new positions of wolves are generated with the help of two operators, known as enriching and hunting operators. However, how well these operators are exploring search space is dependent on random parameters A and C used in defining equations. The correct values of these parameters can improve the convergence rate and produce an optimal global solution for a problem in fewer generations. Deciding the right value of the random parameter is known as parameter tuning. Most of the time, the algorithm sticks to a local optimal solution due to improper values of these parameters. A selection operator is used to mark the best solutions from the entire search space. Therefore selection will be performing exploitation.

The algorithm shows that the values of alpha, beta, and delta generated in the current generation do not affect the next iteration's alpha, beta, and delta solutions. This property makes the algorithm more suitable for solving multimodal problems. Furthermore, to improve the performance of GWO, some changes in operators can be made to create a proper balance between exploration and exploitation.

A proper balance between exploitation and exploration is required to improve the performance of the algorithm.

6.3 DETAILED IMPLEMENTATION OF A GWO PROGRAM

To implement the hunting process of grey wolves in the form of a computer program, we have to define the movement of wolves during their hunting process (searching for prey, enriching of target, attacking the prey) in the form of functions/operators. The GWO algorithm aims to search multiple regions in search space to identify the best solution as early as possible. Two functions/operators, that is, enriching and hunting, are used in the computer program to define the movement of grey wolves during the hunting process. In the program, wolves are represented as solutions in the search space. The change in the position of wolves is defined in several steps. These steps are called iterations. In each iteration, the positions of wolves are updated to get new positions. The method used to update the position of wolves in a particular iteration is shown in the following subsections.

6.3.1 Generating Initial Population

In each iteration, the known positions of wolves are passed to produce new positions for the wolves. No information about wolves' positions is available in the initial iteration, so each wolf's initial position is generated randomly from the given input domain. The initial positions of all wolves define the initial population for the first iteration. Later, when updated positions of wolves are available. They are supplied to get new positions for the wolves. For example, in the first iteration, no information is available about the position of particles. So these values are randomly generated from the given search domain. After completing the first iteration, the output of the first iteration will be supplied as an input to the second iteration. Similarly, the production of the second iteration will be provided as an input to the third iteration.

6.3.2 Identification of Alpha, Beta, Delta, and Omega Solutions

After passing the input solutions, the fitness of each solution is calculated by checking the objective value of the function at each solution. All solutions will be classified into four categories, that is, alpha, beta, delta, and omega, based on fitness values. The solution with minimum fitness is known as alpha. The second minimum is known as beta. The third minimum is known as delta, and all other solutions will be omega. Although alpha, beta, and delta solutions represent the first, second, and third minimum, they may not be the top three solutions. An example will explain this. This happens due to the conditions used in defining the positions of three best solutions. These conditions forced that the position of only one wolf (alpha, beta, or delta) will be updated when the comparison is made between the two solutions. So if more than one solution is used for updating the alpha position of the solution, then the value stored in the second or third best will not be actually the second best and third best. It may be possible that alpha retained it earlier but has replaced it as a new solution is better than the previous position. As there is no provision in GWO for automatically moving those values in beta and delta solutions, this information will become unavailable for other solutions.

6.3.3 Updating Position of Each Wolf and Producing Output

Searching, encircling, and attacking processes are used to update the position of each wolf. The position of each wolf is updated under the leadership of alpha, beta, and delta wolves. Alpha, beta, and delta wolves are used to guide other wolves as they have more information about the position of prey in the pack. Each wolf defines its new position by averaging the tentative positions estimated by alpha, beta, and delta wolves.

To define the new positions of wolves, alpha, beta, and delta wolves estimate the new position of wolves. First, the positions of alpha, beta, and delta are calculated. For calculating alpha, beta, and delta solutions, the fitness of solutions is evaluated. The fitness of each solution is evaluated according to the value of its position vector. The position of the best individual of the current generation is noted as the alpha; similarly, the second- and third-best solutions are known as beta and delta, respectively.

After fixing the values of the alpha, beta, and delta positions, the new position of wolf is calculated.

Mathematically following explanation can be used to define the process of GWO. Let us suppose that that in current population, $\vec{A}$ and $\vec{C}$ are coefficient vectors, $\vec{X}_p$ is the position vector of prey, $\vec{X}$, indicated the position vector of a grey wolf.

The following equations are used for enriching and hunting prey. Finally, a new position of the wolf is obtained by averaging the new positions estimated for the alpha, beta, and delta wolves. The generalized formula for enriching and hunting follows:

$$\vec{D} = \left|\vec{C}.\vec{X}_P(t) - \vec{X}(t)\right| \tag{6.1}$$

$$\vec{D}(t+1) = \vec{X}_p(t) - \vec{A}.\vec{D}, \tag{6.2}$$

where t indicates the current generation, $\vec{A}$ and $\vec{C}$ are coefficient vectors, $\vec{X}_p$ is the position vector of prey, and $\vec{X}$ indicated the position vector of a grey wolf.

The enriching process is used to target the distance between the current position of wolf and the current position of the prey.

$$\vec{D}_\alpha = \left|\vec{C}_1.\vec{X}_\alpha - \vec{X}\right| \tag{6.3}$$

$$\vec{D}_\beta = \left|\vec{C}_2.\vec{X}_\beta - \vec{X}\right| \tag{6.4}$$

$$\vec{D}_\delta = \left|\vec{C}_3.\vec{X}_\delta - \vec{X}\right| \tag{6.5}$$

After the enriching process, the new positions estimated by α, β, and δ wolves are calculated by the following equations:

$$\vec{X}_1 = \vec{X}_\alpha - \vec{A}_1.\left(\vec{D}_\alpha\right) \tag{6.6}$$

$$X_2' = X_\beta' - A_2'.\left(D_\beta'\right) \tag{6.7}$$

$$\vec{X}_3 = \vec{X}_\delta - \vec{A}_3.\left(\vec{D}_\delta\right) \tag{6.8}$$

Finally, new position of the solution is calculated by taking the average of estimated positions.

$$\vec{X}(t+1) = \frac{\vec{X}_1 + \vec{X}_2 + \vec{X}_3}{3} \tag{6.9}$$

The vector $\vec{A}$ and $\vec{C}$ are calculated as follows:

$$\vec{A} = 2.\vec{a}.\vec{r_1} - \vec{a} \tag{6.10}$$

$$\vec{C} = 2.\vec{r_2} \tag{6.11}$$

6.3.4 Clamping

New positions obtained by Equation 6.9 are clamped to the range `[-xmin, xmax]`. This process is used to control excessive roaming of solutions outside the search space. If the value of a variable became out of bound, it will be replaced by the boundary values.

This process is repeated until a user-defined stopping criterion is reached.

Pseudocode of GWO algorithm:

Pseudocode for GWO can be represented as follows, the pseudocodes used in designing the algorithm are listed in Table 6.1.

TABLE 6.1
Notations Used in a GWO Algorithm

Notations	Description
`X`	Current population
`A,C`	Random parameters
`c2`	Constants
`w`	Inertia weight
`Mgen`	Maximum number of generations
`ps`	Population size
`Dim`	Dimension

Algorithm 6.1: Algorithm of GWO

```
GWO()
begin
 x = rand(ps,Dim) // Initialize the position of wolf with a random vector
 evaluate the function cost for each solution
α = solution having minimum fitness in current generation,
β = second minimum of current generation,
δ = third minimum of current generation
// update the swarm's best-known positions
 Do
```

foreach solution in iteration `i = 1 to Mgen` **do**

Calculate:

$$\vec{D}_\alpha = \left|\vec{C}_1 . \vec{X}_\alpha - \vec{X}\right|$$

$$\vec{D}_\beta = \left|\vec{C}_2 . \vec{X}_\beta - \vec{X}\right|$$

$$\vec{D}_\delta = \left|\vec{C}_3 . \vec{X}_\delta - \vec{X}\right|$$

After enriching new position estimated by α, β, and δ wolves are calculated by following equations:

$$\vec{X}_1 = \vec{X}_\alpha - \vec{A}_1 . \left(\vec{D}_\alpha\right)$$

$$\vec{X}_2 = \vec{X}_\beta - \vec{A}_2 . \left(\vec{D}_\beta\right)$$

$$\vec{X}_3 = \vec{X}_\delta - \vec{A}_3 . \left(\vec{D}_\delta\right)$$

New position

$$\vec{X}(t+1) = \frac{\vec{X}_1 + \vec{X}_2 + \vec{X}_3}{3}$$

end

end

6.3.5 Parameter Tuning

Parameter vectors A and C play a very important role in defining the convergence of a GWO algorithm. The performance of a GWO algorithm can be changed by changing the values of these parameters. Changing the performance of the GWO by changing the values of these parameters is known as parameter tuning. The setting of these parameters is dependent on the expert's knowledge about the problem. In some GWO programs, setting these parameters does not require any expert. It is done by the program itself. Such algorithms are called self-adaptive algorithms.

6.4 MATLAB® Implementation of a GWO Program

MATLAB instructions for writing the GWO program is explained in the following subsections.

6.4.1 Defining the Initial Population

The process of initialization in GWO is the same as it is in PSO. In the GWO algorithm, the population is also initialized by randomly selecting solutions from the entire search space. The population size has a significant impact on the algorithm's convergence rate. In most algorithms, the minimum size of the population for solving a complex problem is `40`. In the current example, we used ten solutions in the initial population. In MATLAB, a variable can be declared for fixing the size of the initial population. In this program, the variable name `popsize` is used to define the population size, and it is initialized simply by writing this command `popsize = 10;`

Dimension and fitness function:
After initializing the population, we must work according to the algorithm's theory. In GWO, solutions are represented by the wolves' positions. As a result, to define the initial position of wolves and then update their positions, we must define the initial position and updating mechanism for each solution. The positions of wolves in the first generation are generated at random. The first generation of solutions will define the initial population. Because GWO is used to solve an optimization problem, we want to identify the value of the optimal solution that will produce the optimal value of the function. We have to check how many variables will be used for defining the solution vector. These variables represent the dimension of the solution vector.

Let us take an example and check the dimension of the solution vector.

The function which we want to optimize is known as the fitness function. Let us suppose that we want to optimize (minimize) a function, `F`, which is defined as follows:

$$Minimize\, F\left(x_1, x_2, x_3\right) = x_1^2 + x_2^2 + x_3^2,\ where -5 \le x_i \le 5. \tag{6.12}$$

It is a multivariable function with three variables for each solution. The number of variables in a solution will define the dimension of the solution, and for the given problem, it will be three. This problem's fitness function is a three-variable function whose value is determined by calculating $F\left(x_1, x_2, x_3\right)$. The optimal solution to this problem can be represented by `xopt`, and it will also have three variables. Variables in optimization problems are bounded, with lower and upper bounds. The lower and upper bounds of each variable in this problem are `-5` and `5`, respectively. It may be different in other cases.

Defining boundary vectors:
In our example, we have a three-variable function with lower and upper bounds of `-5` and `5`, respectively. A minimum vector, also known as an `xmin` solution, is commonly used to define a solution in which each variable is represented by the lower bound. Similarly, `xmax` is known as the maximum vector, in which all variables are at their upper bounds. In MATLAB, we can define the minimum and maximum vectors using the following steps.

These vectors are used to create boundary vectors for both position and velocity, as we can see. After defining these variables, we can use a MATLAB command `ones(row,column)` to generate a matrix of `row x column` in which the value of all variables will be `1`. The `row` variable specifies the number of rows, and `column` specifies the number of columns.

For example, let us suppose that we want to generate a three-variable vector in which all variables are `1`. Then this can be done by writing the following command.

```
ones(1,3) = {1, 1, 1}
```

After generating this vector, we can multiply it with `xbound = 5` and `vbound = 5` to create maximum and minimum vectors.

For example

```
Xmin=- xbound*ones(1,DIM)
Xmax= xbound*ones(1,DIM)
```

Population initialization:

The maximum and minimum vectors are generated. We can develop the initial population. Because the size of the initial population is ten, we must generate ten positions. These vectors will be generated within the given variable boundaries.

In MATLAB, we can generate random values for all solutions in the form of a matrix with a single statement `rand(row, column)`. However, these random values will be generated between (`0` and `1`) in our problem, where 0 is the lower bound and 1 is the upper bound.

We have to generate values in the interval `(-100,100)`.

This can be generated as follows:

First generation:

In the GWO algorithm, the initial values of the `alpha`, `beta`, and `delta` solution vectors are `(0, 0, 0)` and their fitness is set to infinite.

Following the fixation of these values, the initial population is generated within the given bounds.

Each variable in the current example is bounded between `-100` and `100`.

We can use a variety of methods to generate ten random solutions within these constraints.

One method of population production is depicted in the following.

We begin by creating a temporary matrix.

```
t1 = rand(popsize,DIM)
```

And then following equation can be used to generate values within the given bounds.

```
x = 2 * xbound * t1 - xbound
```

Let us suppose we want to define the values of ten solution vectors in which three variables represent each solution. Then we have to make `popsize` 10 and `DIM` 3, and the following command will be used `rand(popsize,DIM)`

6.4.2 Updating Alpha, Beta, and Delta Solutions

The following equations are used to update the `alpha`, `beta`, and `delta` positions. When the entire population has been generated, each solution's fitness is calculated. Following that, the following equations are used to update the values of `alpha`, `beta`, and `delta` solutions.

TABLE 6.2
Initial Population

60.069	6.977	-85.3171
62.1349	-23.0311	10.5587
-64.2899	-24.9548	-16.6621
38.8587	94.5839	-55.1052
32.1954	42.615	-4.3544
98.0125	-30.8218	-0.139
-23.297	-51.249	20.5827
46.101	51.1316	-42.8395
-84.7601	76.9769	7.9141
-85.0748	-30.6854	3.9025

Starting with the first solution, the previous fitness of the alpha solution is compared to the fitness of the new solution.

If the new fitness is less than the previous solution, the alpha solution will be updated.

For example, consider the first solution (60.0690, 6.9770, -85.3171) and its fitness of 8.4373e+03. The condition for updating the alpha solution is checked

fitness<Alpha _ score since 8.4373e+03 is less than infinite.

Therefore, the alpha score will be updated to 8.4373e+03, and then the alpha solution is also updated to (60.0690, 6.9770, -85.3171). Nothing will be done to the beta and delta positions. These solutions and their score will be updated only when the alpha score is not updated. Now fitness of next solution is checked. If the new fitness is less than alpha score, the value of the alpha score will be updated again. The fitness of solution (62.1349 -23.0311 10.5587) is 7.8562e+03, which is less than 8.4373e+03. Again, the alpha score is updated to 7.8562e+03 and the alpha position to (62.1349 -23.0311 10.5587). No change will be done in the beta and delta solutions.

A similar process will be used for other solutions for updating the alpha, beta, and delta solutions and their scores.

Now the fitness of the third solution (-64.2899 -24.9548 -16.6621) is calculated. This fitness, 2.3314e+04, is greater as compared with existing alpha score, which is 7.8562e+03.

Now the alpha score will not be updated as the new fitness is greater than the previous alpha score, but other conditions are checked to see whether there will be some change in the beta or delta solution. The first chance will be given to the beta, and the following condition is checked:

fitness>Alpha _ score && fitness<Beta _ score update Beta _ score and Beta _ solution

Although the fitness of the third solution is greater than the alpha, yet it is better than the previous value of the beta score; therefore, the beta score will be updated to 2.3314e+04, and the beta position will be updated to (-64.2899 -24.9548 -16.6621).

Similarly, the fitness of the fourth solution is checked to see whether there will be changes in the alpha score or the beta score. As the fitness of this new solution (38.8587 94.5839 -55.1052) is 2.5454e+04 greater than the alpha and beta scores, no change will be made in the alpha and beta scores. However, a third condition is checked to see if the delta score is updated.

The third condition is fitness>Alpha _ score && fitness>Beta _ score && fitness<Delta _ score

According to the third equation, the delta score, along with position, will be changed to a new value. The new delta score will be updated to 2.5454e+04, and the delta solution will be updated to (38.8587 94.5839 -55.1052).

After this solution, the fitness of the fifth solution (32.1954 42.6150 -4.3544) is evaluated and used to update the previous scores of each alpha, beta, and delta.

The fitness of (32.1954 42.6150 -4.3544) is greater than the alpha score but less than beta score. So the second condition will be satisfied, and new value for the beta score will be calculated.

The new value of the beta score, along with its position, is mentioned in the following:

Beta position (32.1954 42.6150 -4.3544)

Beta score 1.1597e+04

The fitness of the sixth solution (98.0125 -30.8218 -0.1390) is 1.8617e+04, which is worse than the previous alpha and beta scores but better than the previous delta score, so the delta score will be changed, and the delta solution is also updated.

Delta position (98.0125 -30.8218 -0.1390)

Delta score 1.8617e+04

The fitness of the seventh solution (-23.2970 -51.2490 20.5827) is 9.0119e+03. This new fitness is worse than the previous alpha score, but it is better than the beta score. The new beta position and score will be as follows:

Beta position (-23.2970 -51.2490 20.5827)

Beta score 9.0119e+03

The fitness of the eighth solution (46.1010 51.1316 -42.8395) is 1.4538e+04. This fitness is better than the delta score, so the delta score will be updated. The new delta score and position will be as follows:

Delta position (46.1010 51.1316 -42.8395)

Delta score 1.4538e+04

Similarly, the fitness 7.2449e+03 of the ninth solution (-84.7601 76.9769 7.9141) is even better than the previous alpha score.

New alpha score and solution will be as follows:

Alpha position (-84.7601 76.9769 7.9141)

Alpha score 7.2449e+03

The fitness, 3.3150e+04, of the tenth solution (-85.0748 -30.6854 3.9025) is worse than all alpha, beta and delta scores. Hence, no change will be made in the previous values of alpha, beta, and delta. The final alpha, beta, and delta solutions, along with their fitness, are highlighted in the following table.

Bold values show the `alpha`, `beta`, and `delta` solutions for the current generation.

Solutions			Function Cost	Alpha, Beta, and Delta Solutions
60.0690	6.9770	-85.3171	8.4373e+03	
62.1349	-23.0311	10.5587	7.8562e+03	
-64.2899	-24.9548	-16.6621	2.3314e+04	
38.8587	94.5839	-55.1052	2.5454e+04	
32.1954	42.6150	-4.3544	1.1597e+04	
98.0125	-30.8218	-0.1390	1.8617e+04	
-23.2970	**-51.2490**	**20.5827**	**9.0119e+03**	**Beta**
46.1010	**51.1316**	**-42.8395**	**1.4538e+04**	**Delta**
-84.7601	**76.9769**	**7.9141**	**7.2449e+03**	**Alpha**
-85.0748	-30.6854	3.9025	3.3150e+04	

Updating positions:
These values of `alpha`, `beta`, and `delta` solutions are used to update the positions of solutions in each dimension. For updating the position of each solution in a given dimension, the average of the new position values estimated by three wolves are used. The new position of each wolf in a given dimension is estimated by the three wolves using the equation for enriching the wolf and then the equation for attacking on the wolves; see Equations 6.1 and 6.2. Enriching the prey is done with the help of D value estimated by the various wolves. Finally, these values of D are used to update new position of wolf in a given dimension.

For example, for updating the position of the first wolf in the first dimension, the first estimated value by alpha wolf is calculated and is used to calculate new position. Similarly, the other values will be calculated by the beta and delta wolves, and the average of all values are taken. This value will be the new position of the first dimension for the first wolf.

Mathematically, the following steps are used to update the first dimension of the first wolf. A similar method will be used to update the other dimensions of same solution.

First, we fix a for all equations. We have fixed `a` as 2.

```
a = 2
```

Now we will generate two random values and estimate the values of A and C. Update the position of search agents, including omegas.

```
r1 = 0.4782
r2 = 0.5810
A1 = 2*r1*a-a = 2*2*4782-2 = 1.9128-2 = -0.0872
C1 = 2*r2 = 1.1619
```

For calculating D for `alpha` using Equation 6.3, we have to check the first dimension of `alpha`, which is `-84.7601`, and the value of the first solution in the first dimension, which is 60.0690. After putting these values into the enriching equation, we get

```
absolute(1.1619*-84.7601-60.0690)= 158.5532
```

This value of D will be used to estimate the new value for the alpha solution. As per the formula of Equation 6.6, we get

```
-84.7601-(-0.0872*158.5532)=-84.7601+13.8258
X1 = -70.9288
```

Similar to these calculations, the same calculations will be done for beta and delta solutions. The estimated values X2 and X3 are calculated. Finally, the updated position of the first solution in the first dimension will be calculated by taking the average of X1, X2 and X3. The calculation of X2 and X3 is shown.

```
r1 = 0.3908
r2 = 0.9206
A2 = -0.4367
C2 = 1.8411
D_beta = 102.9617
X2 = 21.6619
r1 = 0.6764
r2 = 0.2915
A3 = 0.7056
C3 = 0.5829
D_delta = 33.1952
X3 = 22.6798
```

After these steps, the first dimension of the first solution is updated. For updating other solutions, other values of required parameters are generated. This process is repeated for each dimension and each solution, and new positions are created. This process will be repeated again and again until all solutions update their positions.

Positions =

-8.8624	6.977	-85.3171
62.1349	-23.0311	10.5587
-64.2899	-24.9548	-16.6621
38.8587	94.5839	-55.1052
32.1954	42.615	-4.3544
98.0125	-30.8218	-0.139
-23.297	-51.249	20.5827
46.101	51.1316	-42.8395
-84.7601	76.9769	7.9141
-85.0748	-30.6854	3.9025

These solutions are again updated in new iterations by choosing new values of alpha, beta, and delta.

7 Environmental Adaptation Method

7.1 INTRODUCTION

Evolutionary learning has been used in the design of various algorithms in evolutionary algorithms. Evolutionary learning describes how the genetic structure of an offspring population changes as a result of selection, crossover, and mutation in the parent population. According to the theory of survival of the fittest, these genetic changes make the offspring population fitter than the parent population. These changes in the genetic structure of the solutions lead to an optimal genetic structure generation after generation. This is the primary reason why evolutionary learning is used repeatedly to solve optimization problems. We have already seen how genetic algorithms (GAs) and differential evolution algorithms work and how evolutionary learning can be used to solve optimization problems. Although there are many evolutionary algorithms in the literature, scientists are still updating existing algorithms or developing new ones that can overcome the shortcomings of existing optimization algorithms. The primary goal of these new algorithms is to either increase the algorithm's convergence rate or to generate more efficient solutions to extremely complex optimization problems. This chapter goes over a new algorithm called the environmental adaptation method. Before delving into the algorithm, we go over the issues with the existing evolutionary algorithms and then show how the performance of the Environmental Adaptation Method (EAM) has been improved.

Existing optimization algorithms have two significant limitations: slow convergence rate and stagnation problem. To create a fast algorithm, we must first identify the variables on which these parameters are dependent. We can make changes and build new algorithms once we have information about these factors. The algorithm's convergence rate will be slow compared to other algorithms in which a different natural phenomenon is used, which takes less time in producing an offspring population. In general, the evolutionary process takes time as changes in genetic structure will be observed in the offspring population only when evaluated by other peers. To improve the convergence rate of evolutionary algorithms, we should use a different phenomenon that takes less time. Adaptive learning is a component of the evolutionary learning process. It takes less time to be a subprocess of evolutionary adaptive learning. As a result, adaptive learning was used to design a new algorithm to create an optimization algorithm with an improved convergence rate. This new algorithm is known as EAM. It was assumed that it would produce reasonable solutions in a shorter period.

DOI: 10.1201/9781003313649-7

7.2 EAM

EAM is a novel evolutionary algorithm that applies adaptive learning to optimization problems. It is a population-based algorithm, as are other evolutionary algorithms. At first, a binary version of EAM is proposed. Later, real-parameter versions are presented as well.

7.2.1 Natural Phenomena Used for Designing EAM

There are two kinds of changes that may occur in species to make them more suitable in a given environment. The first type of change takes place in the genetic structure of the species. These changes take time and are permanent.

Phenotypic changes are another type of change that influences an individual's fitness. These changes are distinct from genetic changes. These modifications define how an individual's behavior changes in a given environment. These changes are measured by observing changes in a species' phenotypic structure (which explains the behavior of species at a given time). As time passes, species try to improve their phenotypic structure and update iteratively until it is optimal. According to the theory of adaptive learning, changes in the phenotypic structure of species occur.

This theory states that if a population (with plastic traits) finds itself in a new environment where its individuals' phenotypes are not optimal, suboptimal individuals will acquire higher fitness due to adaptive plasticity. This learning process is repeated repeatedly until the population attains optimal fitness EAM implements Baldwin's theory of adaptive learning.

7.3 DETAILED DESCRIPTION OF AN EAM PROGRAM

There are two versions of EAM that are widely used. There are two types of EAM: binary EAM and real-parameter EAM. Although the theory of adaptive learning is well defined, the implementation of EAM operators is still evolving. The designs suggested in the following sections are likely to change in future versions. The first version of EAM was known as binary EAM because the algorithm was designed using the binary encoding of the solution. Let's take a look at the algorithm's implementation.

7.3.1 Binary EAM

The first version employs binary encoding. The adaptive learning process has been divided into three steps for writing the binary version of EAM: adaptation, alteration, and selection. These steps employ a binary representation of the solutions that represent an individual's phenotypic structure. This algorithm, like other population-based algorithms, begins with a randomly generated initial population.

7.3.1.1 Adaptation Operator

During the adaptation step, the phenotypic structure of randomly generated solutions is updated to form a more suitable phenotypic structure. These changes in phenotypic structure are implemented using binary string decimal values. Each solution's phenotypic structure is updated to a new decimal value within the given bounds by using Equation 7.1, where its fitness and environmental fitness guide each solution. This conversion is accomplished by assessing the current ecological and individual fitness and selecting two random adaptation random parameters. The equation for calculating the new decimal value of a binary structure is calculated using Equation 7.1.

The decimal value calculated by Equation 7.1 is then converted into a new binary string that defines a new phenotypic structure. The equation used for determining phenotypic conversion is shown:

$$P_{i+1} = \left[round\left(\alpha(g) * (P_i)^{\frac{F_n(P_{in})}{F_{avg}}} + \beta(g) \right) \right] \%2^L \tag{7.1}$$

$\alpha(g)$ and $\beta(g)$ are random numbers used in gth generation and these values are decided as per the requirement of the problem, L represents the total number of bits in an individual, F_avg is the average fitness value of the current population, representing the current environmental fitness and g is the generation number. After adaptation, each solution is represented again, represented in binary.

Let us discuss why this equation is suitable for defining the search in the correct direction.

Let us look at this problem and try to find the best solution.

Assume we are using binary encoding and have already divided the given search space into 256 solutions to find the best solution. Because binary encoding is used, these solutions can be represented by index values ranging from 0 to 255. A decimal value can be used to represent any solution, global or local. To begin any population-based algorithm, we start with some random solutions that serve as the initial population. These solutions are denoted by the letters A, B, C, D, E, F, and G here. In other algorithms like GAs, the selection is first applied to identify which solutions may produce optimal global solutions. Clearly, during selection, A, C, and F will be rejected as they will never produce a global optimal solution. The selected solution will be B, D, and G. Following selection, the area around these solutions is used to identify other possible solutions. Simultaneously, some exploration is carried out in order to identify other possible regions where a global optimal solution may exist. How does this exploitation and exploration take place? In general, some solutions near selected solutions are generated by adding or subtracting some integer values from previous solutions. If the value to be added or subtracted is less, the newly defined solution will perform exploitation. If it's large, they'll go exploring. In a GA, this addition or subtraction is carried out in the crossover and mutation operators.

In the crossover operator, the exploitation will occur when the cut point is near the least significant bit, and the exploration will take place when the cut point is close to the most significant bit. Similarly, mutation exploitation will occur when the bit flipping is done around the least significant bit and the exploration is done when bit flipping is done close to the most significant bit.

To clarify, consider the following example. Let's try it on two binary strings: 10001110 and 01100001. Let's pick a cross-site near bit position 1. Then newly generated solutions will be

Parent solution's offspring

100011 10 10001101
011000 01 01100010

When you compare the parent and offspring, you will notice that the newly generated solution 10001101 is one step ahead of its parent 10001110. So 10001101 is obtained by subtracting −1 from its parent. Similarly, 01100010 is one step back from its parent 01100001 and can be obtained by adding +1 to the previous value.

Similarly, we can look for the mutation operator. It is clear that to generate new solutions, some processes must be implemented in any optimization algorithm. We did things a little differently in EAM.

Because adaptation in EAM changes the phenotypic structure of any solution to an unknown value. We took decimal values of solutions and converted them into other decimal values to make this process possible. Because newly generated values may be floating or out of bounds. The first equation generates a new floating-point value and converts it to a given bound using the modulus operator. Finally, the floating value is converted to an integer value using the round function. This explains how the equation is designed for the adaptation operator.

7.3.1.2 Alteration Operator

Look some changes in the phenotypic structure of the solutions may occur during alteration due to environmental noise that may modify one or more bits of the solutions. Not all solutions will be altered; it will depend on the likelihood of change. The alteration operator is used to implement these changes. However, it appears to be a mutation operator used in GAs. New alteration operators are being developed to improve the convergence rate.

7.3.1.3 Selection Operator

An intermediate population is formed by new structures (solutions) that emerge as a result of adaptation and modification. The initial structures are then combined with the intermediate structures for selection. Those structures that survive in a new environment are used to create a new mean of fitness. Structures that do not contribute to the new environment are destroyed.

These three steps will be repeated until either the solution with the desired fitness is captured or the process reaches its maximum number of iterations.

Because these changes occur over the course of a single life, the possibility of finding a good solution in a short time is high.

7.3.1.4 Explaining EAM with One Numerical Example

The details of EAM follow. The operators are named adaptation, alteration, and selection operators.

Adaptation: An adaptation operator implements the changes in phenotypic structure in each solution by taking guidelines from the current and overall average fitness of all solutions. In the absence of any other information, this average fitness defines the current environment. An updated phenotypic structure of each solution P_i can be identified as follows:

$$P_{i+1} = \left[round\left(\alpha(g) * (P_i)^{\frac{F_n(P_{in})}{F_{avg}}} + \beta(g) \right) \right] \% \, 2^L \tag{7.2}$$

This equation changes the phenotypic structure of each solution, where P_i is the decimal value of binary-coded *i* th solution and 2^L is the total number of solutions. F_n (P_{in}) The decoded value in the decimal of the binary coding of P_{in} is the fitness value of P_{in}; $\alpha(g)$ and $\beta(g)$ are random numbers used in *g* th generation that are decided as per the requirement of the problem; *L* represents the total number of bits in an individual; F_{avg} is the average fitness value of the current population, representing the current environmental fitness; and *g* is the generation number? The % operator is used in EAM for clamping purposes. The values of a $\alpha(g)$ and $\beta(g)$ are updated in each generation, and this value is dependent on the value of fitness of the solution. For good solutions, this will be less so that exploitation can be performed. For solutions having fitness less than average, the value will be significant.

Alteration: Although adaptation works on decimal numbers, alteration directly updates binary coding. An alteration operator generates a new solution P_{i+1} by flipping one or more bits of P_i.

Selection: The best solutions equal to the initial population are selected from a combination of the current and previous generations to form the current generation.

Let us take an example and see how the operators are working in EAM and other versions. Let us suppose we want to optimize a function *f (x)*, which is defined as follows:

$$Min\, f(x) = \frac{x}{1+x^2}, \; where \;\; 0 \le x \le 3. \tag{7.3}$$

For representing solutions in binary, the search space will be divided into 2^1 number of solutions. If we divide the search space into 64 solutions, then each solution can be represented by 6 bits. To create the initial population, we start with some random solutions generated from the whole search space. Let the binary representation of these solutions be `000001,010010, 001110, 101000`. This randomly generated solution will be supplied as input and updated according to Equation 7.1. Since Equation 7.1 is working with the decimal value, the decimal equivalent of each binary string is shown in Figure 7.1. Finally, the third value is delivering actual value.

The first adaptation operator is applied, and every solution will be updated according to Equation 7.1. Let us see how solution `0000001` will be adapted. For using Equation 7.1, the value of α(g) and β(g) should be generated randomly. The value of α(g) and β(g) should be generated in closed interval `[0, 2`^{1} `-1]` and `[0,1]`

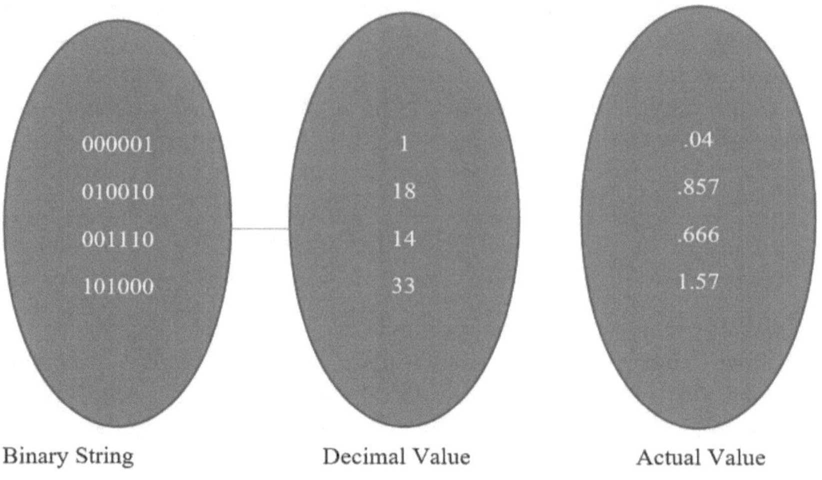

FIGURE 7.1 Binary—Decimal and the Actual Representation.

respectively. Let us suppose these values are `3` and `.5` and the decimal value for the given binary string will be `1`; the actual value will be `1/2`^{1} and 2^1 `-1` will be `63`. Similarly, the value of $F_n(P_{in})$ i.e., the decoded value in decimal of the binary coding of P_{in} is `.04/1+ (.04)`^{2}`=.0399`. Similarly, the fitness values of other solutions will be `4941, .4613,` and `.4531`. The average fitness of all these solutions will act as environment fitness equal to `.3621`. Now we will calculate the new value of the first solution; for that, first we calculate `0399/.3621=.1101`

So new value of solution one will be:

`= round (3(1)`$^{.1101}$`+.5)%64=round(3.5)%64 = 4`. Since 4 is less than `63`, no clamping will be required. This 4 will be the decimal value for the binary representation. Hence, the binary value for this decimal number will be `0001000`. This is the binary value where 4 is represented in 6-bit.

Similarly, for the second solution, new decimal and binary values will be calculated as follows:

`round (3*(18)`$^{1.364}$`)+.5)=round(155.1)%64=27`

Similarly, new values of third solutions can be calculated.

`round (3*(14)`$^{1.251}$`)+.5)=round(238.6)% 64 =239%64=47`

Similarly, new value of fourth solutions can be calculated.

`round (3*(33)`$^{1.273}$`)+.5)=round(86.8)% 64 =23`

So intermediate population after adaptation will be as shown in Figure 7.2.

Alteration operator: The alteration operator is intended to change one or more bits of the new phenotypic structure. This was accomplished by first checking a random value between `1` and the maximum length of the string. This value will determine which half of this value will be used. After calculating the middle position of the bit

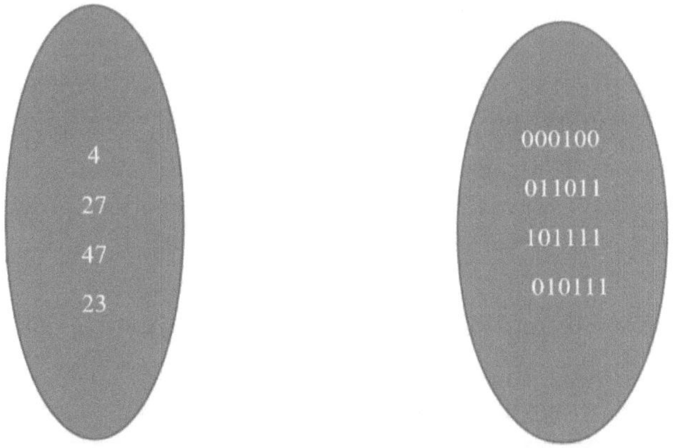

FIGURE 7.2 After adaptation, new phenotypic structures of solutions.

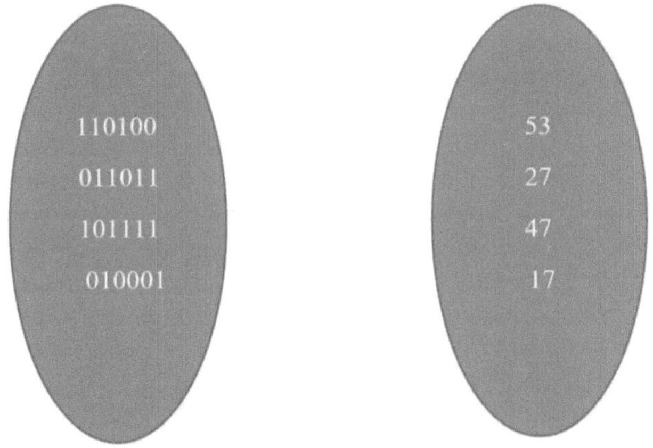

FIGURE 7.3 Solutions after alteration operator.

string, the newly generated value will be compared to the mean value. If the newly generated value is in the upper half, the upper half's bits will be flipped. If it is found in the lower portion, the bits in the lower portion will be flipped. Not all solutions will perform a modification. To modify the solutions, an alteration probability was used. Assume that if the probability of change is `.5`, then `4*.5` solutions will fall under the category of change. As a result, two solutions will be altered. First, two random solutions are chosen from the adapted solutions, and then alteration is used. Assume that the two solutions to be changed are `4` and `23`. Then two more random integer values are generated. If the numbers are `5` and `2`, The average value of string position will be `2.5`. So, in one solution, bits `3` to `5` will be chosen at random. Three random bits will be generated at random. If these are `110`, the first solution after modification will be `110100`. Lower bits will be changed for the second solution. Assume these values are `001`. The new solution after the change will be `010001`. As a result, a new population will emerge as a result of adaptation and modification.

These values will define the intermediate population and combine with the parent population for performing selection.

Selection operator: The population generated after adaptation and alteration is merged with the input population, and the four best solutions are selected as new output solutions.

7.3.2 Understanding the Relevance of Phenomena with Optimization

The EAM process for solving optimization problems is very different from existing algorithms. The first selection operator is used in genetic algorithms to find good solutions in the search space. These selected regions are exploited and explored to find other good solutions in the search space. Unlike GAs, EAM first employs adaptation and alteration operators to explore and exploit the entire search space. EAM's adaptation operator is designed to exploit the region surrounding good solutions while also exploring solutions from the other areas. This exploitation and exploration are implemented with the help of random parameters $\alpha(g)$ and $\beta(g)$. The parameter $\alpha(g)$ is responsible for guiding solutions in new search areas and exploration. A large value of $\alpha(g)$ performs a thorough investigation of the search space. $\beta(g)$ is used for exploiting the regions and avoiding sticking on local optima. The new solutions generated after adaptation and alteration are then combined with the input population and passed in the selection operator. Selection will keep all good solutions and discard bad solutions. In prior iterations, the value of $\alpha(g)$ and $\beta(g)$ are kept large so that search space can be adequately explored. In later generations, the value of $\alpha(g)$ and $\beta(g)$ are decreased uniformly so that solutions can be converged on global optimal solutions. Similar to EAM, adaptation is applied to every solution, and a new population is created. This newly created population is then combined with the previous population for selection. Finally, selected solutions of this generation act as input to the next generation.

Because the process is more focused on exploration in the first few iterations, in unimodal problems, the EAM convergence rate is low. However, it performs well on multimodal and shifted functions.

7.3.3 Real-Parameter EAM

The EAM process used in the binary version is complex. For real-valued problems, a real-parameter version of EAM is recommended for creating simplified versions. In this version, two modifications are suggested to keep things simple. Instead of three operators (adaptation, alteration, selection) in real-parameter version, only two (adaptation and selection) are used. It was thought that adaptation operators could replace the work of adaptation and alteration operators. Furthermore, the complex equation used in binary EAM has been simplified in the new version. The new equation, Equation 7.4, used in real-parameter EAM is shown:

$$P_{i+1} = \left(c(g) * (P_i) + \beta(g)\right), \tag{7.4}$$

where $c(g)$ is equal to a(g)*$F_n(P_{in})/F\_avg$.

The convergence rate of the new real-parameter version is greater than the previous version. The reason for improved convergence rate lies in the new operator with a better exploitation capability than the previous version.

7.4 IMPROVED ENVIRONMENTAL ADAPTATION METHOD

The motivation for developing EAM was to improve the algorithm's convergence rate and make it more efficient in dealing with multimodal problems. We saw that the performance of EAM was good in solving multimodal problems, but the convergence rate of EAM was not as good as expected. The EAM convergence rate was poor because the early-stage adaptation operator performed very little exploitation. Some changes are made to the EAM adaptation operator in order to improve the EAM's convergence rate. This modified operator was capable of exploitation in both early and late stages. This adaptation operator change is carried out with the assistance of good solutions identified during the selection operator. The updated version performs not only random exploration but also exploitation around good solutions.

The Improved Environmental Adaptation Method (IEAM), like EAM, has two versions. One works with binary encoding of the solution, while the other works with real-parameter encoding

7.4.1 Binary Version of IEAM

The binary version of IEAM follows the same steps as the EAM version. The only operator changed in IEAM was adaptation. Unlike EAM, two equations were used to update the phenotypic structures of the solutions. One equation was the same as it was in EAM. This type of learning, however, was only used to find the best solution. All other solutions were directed toward the current best solution to obtain a new structure that was close to the best solution. In this manner, they were instructed to conduct exploitation around the best solution. Using Equations 7.5, the exploitation of the best solutions is carried out.

For the best solution, the following adaptation operator is used:

$$P_{i+1} = round\ ((\alpha(g) * (P_i)^{\frac{F(X_i)}{F_{avg}}} + \beta(g))\ \%\ 2^L) \tag{7.5}$$

For all other solutions, the following equation is used:

$$P_{i+1} = round\ ((\alpha(g) * (P_i)^{\frac{F(X_i)}{F_{avg}}} + \beta(g)(G_i - P_i))\ \%\ 2^L) \tag{7.6}$$

G_i represents the decimal equivalent of the best solution of the current iteration.

7.4.2 Real-Parameter IEAM

Real-parameter algorithm implements the same equations and operators as the Real version of EAM. Only one change has been made to the adaptation operator.

To learn, two equations are used instead of one. One for the best solution and another for the least desirable. The new modified operator used in the IEAM real-parameter version is

For best solution, the following adaptation operator is used:

$$P_{i+1} = \left(c(g) * (P_i) + \beta(g)\right) \tag{7.7}$$

For all other solutions,

$$P_{i+1} = \left((P_i) + \beta(g) * (G_i - GWi)\right), \tag{7.8}$$

where G_i $and GWi$ are used for representing the phenotypic structure of the best solution of the current generation and the worst solution of the same generation.

These updated versions have a higher convergence rate than EAM, but their efficiency in solving multimodal optimization problems has been reduced. Finally, some changes were made to improve the algorithm's effectiveness in solving both types of problems.

7.5 MATLAB® IMPLEMENTATION OF EAM AND IEAM

EAM and IEAM, like other optimization algorithms, are simple to implement in MATLAB. Let us look at the EAM (binary version) program and discuss the meaning of the instructions used to write the program. Because no mutation has been implemented in this program, it is a variant of EAM.

7.5.1 Generating Binary Solutions

The first step in the EAM program will be to define the initial population by randomly selecting binary solutions from the entire search space and then converting them to real values.

First, we must determine how many solutions we must employ in defining the population. Assume we have ten solutions. The population size can then be explained by the `ps` variable and set to ten.

```
ps = 10
```

It can be used to define ten random solutions. In binary-encoded programs, we must generate binary solutions. Before developing a binary solution, we need to know how many variables will be present in the solution. `Dim` denotes the number of variables. The number of variables is represented by dim. Let us suppose that if only three variables are there in the solution, then we declare `dim` as `3`.

```
dim = 3
```

After determining the population size and the number of variables, we can define ten binary solutions.

However, we must consider how many bits will be required for each variable. We can define these bits in a separate variable `strlen.` In this case, we will use 3.

```
strlen = 3
```

If a 3-bit string represents each variable, then the following command can be used to generate ten solutions, each with three variables represented by 3 bits.

`pop=round(rand(ps,Dim*strlen));` This `pop` will populate a population of `10` binary solutions, with each variable represented by three bits.

Let us see how this goes.

`rand(ps,Dim*strlen)=rand(10,3*3)=rand(10,9).` it will generate 10 rows of 9 random values.

To simplify the problem, suppose `ps=2` and `Dim=3` and `strlen=2`, then `rand(3,6)` matrix will be generated=`[.1.3.4.6.7.1,.5.6.3.7.8 .2]`? When the round is applied to this matrix, all values greater than `.5` are converted to `1`, and all values less than. `5` are converted to `0`.

So the binary matrix will be `[000110, 110110]`.

Converting binary solutions into decimal and real-valued solutions:
This is how the initial population will be generated. After generating binary values, we must convert these solutions to real and decimal values. We can determine the fitness of the solutions after converting them into real value. Assume that each input variable is bounded by two values, `a` and `b`. The matrix can then be converted into decimal and real values in two lines.

Code for converting binary values into decimal and real values:

```
format long
for i=1:Dim
Dec(:,i)=bi2de(pop(:,(i-1)*strlen+1:i*strlen));
act(:,i)=((b-a)/(2^strlen-1)*(Dec(:,i))+a);
end
```

The first statement of this code format is used to represent real values up to `8` decimal places. The second statement is a `for` statement, which is used to repeat the process. The whole number is divided into several segments and then converted into a decimal or real number.

With this code, we check the binary value of each variable and then convert it to decimal.

For example, the first variable of each binary string will be represented by `[00,11]`, and the `Dec` value of the first variable will be `[0,3]`.

If `a=0` and `b=3`, then the act value is `[0,3]`. We will repeat this process for each input variable.

Finally, the `Dec` values of the solution will be `[0,1,2; 3,1,2]` and the actual values will be

```
Act =[0,1,2;3,1,2]
```

Calculating fitness values:
After calculating actual values, the fitness of each solution can be checked by writing this command.

```
fitness=feval(fun,act');
```

This program is written for COCO BBOB framework, where this formula can calculate the solution's fitness.

Question 1: How will you generate a binary solution for function f(x1,x2,x3)=x1^2+ x2^2+ x3^2. Assume that each variable is bounded by a and b where a=0 and b=3.

Moreover, take a 4-bit string for representing one variable. Write MATLAB command also.

Answer: We can initialize a population of the binary solution by fixing the population size and dimension of the solution.

Let us assume the population size is three and the dimension is also 3. Strlen is also 3.

The population can be initialized by pop=round(rand(3,4*3))

This will you three binary solutions in which each variable is represented by 4 bits.

Question 2: How will you convert a binary solution (101001010000) in to real solution (for function f(x1,x2,x3)=x1^2+ x2^2+ x3^2. Assume that each variable is bounded by a and b where a=0 and b=3.

Answer: Divide 101001010000 according to the bits utilized per variable. After dividing solution will be converted in to 3 variables =(2,1,0)

7.5.2 Solutions Generated after the Adaption Operator

Following the initialization of the population, the adaptation operator will be used to generate a unique solution. We will use the decimal values of solutions for adaptation.

The method for implementing adaptation follows.

As is obvious, adaptation will be performed in MATLAB.

Let us first look at how a function can be defined in MATLAB.

In MATLAB, you can declare an adaptation function by writing the function name with input arguments on the left side. On the right, there is a function keyword and an output variable.

```
Function[new_pop,fitad1]=
 adaption(oldpop,old_pop,ps,strlen,Dim,c1,d1,fitness,iter,fa-
vgsel1,Dec)
```

Inside the function, you have to write following commands:

```
c=repmat(fitad,1,Dim);
   favg1=favg*ones(ps,Dim); (this will calculate fi/favg)
```

```
c=c./favg1;
E=favg;
A=randi(0,2^strlen-1)
alpha= a*rand(ps,Dim);
```

`z=alpha.*(oldpop.^c);` (This is for implementing equation used for EAM.)

```
beta=rand(ps,Dim);
```

Look, only decimal values are passed in adaptation.

Finally these values are passed in this equation for converting into another decimal value

```
t=mod((z+beta),2^strlen);
```

This equation will generate another vector of decimal numbers that will be appropriately bounded.

We store these values of t in the new population vector.

```
new_pop=t;
```

Following the adaptation operator, a new population will be generated. This population will be converted into numerical values.

This code explains how the selection operator will be implemented.

7.5.3 Selection Operator in IEAM

First, we define the function, and then we write all the necessary instructions in it.

```
function[sel_pop,fitsel,favgsel]=selection(oldpop,new1_pop,ps,
fun,Dim,b,a,strlen,fitad2)
for i=1:Dim
actm(:,i)=((b-a)/(2^strlen-1)*(new_pop(:,i)))+a;
end
```

This will convert all decimal values to actual values. Finally, these new values with their fitness values will be combined with old values and their fitness values. The best solution will be picked for adaptation.

```
x=fitad2;
f1=cat(2,oldpop,x);
x1=feval(fun,actm');
fin=cat(2,new1_pop,x1');
marge=cat(1,f1,fin);
marge=unique(marge,'rows');
final_sort=sortrows(marge,Dim+1);
sel_pop=final_sort(1:ps,1:Dim+1);
fitsel=sel_pop(:,Dim+1);
favgsel=mean(fitsel);
sel_pop=sel_pop(1:ps,1:Dim);
```

This is how the whole process is used, and this process will be repeated for some fixed number of iterations.

By slightly modifying it, we can create a binary version of the IEAM algorithm.

For running this algorithm 100 times:

```
for j=2:iter

[new_pop,fitad1]=adaption(oldpop,old_pop,ps,
strlen,Dim,c1,d1,fitness,iter,favgsel1,Dec);
fitad2=fitad1;
new1_pop=new_pop;

[sel_pop,fitsel,favgsel]=selection(oldpop,new1_pop,ps,
fun,Dim,b,a,strlen,fitad2);
oldpop=sel_pop;
newpop=sel_pop;
%disp('-----------------------------------------------');
favgsel1=favgsel;
fitness=fitsel;

if fbest > fitness(1)      % keep best
 fbest = fitness(1);
 %xbest = par(1,:);
end
if feval(FUN, 'fbest') < ftarget  % COCO-task achieved
 break;
end
end
```

This segment will run this program `iter` number of times. If `iter` is 100, then this program will run for 100 number of times.

Writing MATLAB program of IEAM:
In the previous section, we saw that we could write the EAM program in MATLAB. We can use similar instructions for a writing program for IEAM (binary version). The only change will be there in EAM and IEAM, and that will be done in the adaption operator of IEAM.

7.6 WRITING MATLAB PROGRAM FOR A REAL-PARAMETER VERSION OF IEAM

Real-parameter versions of IEAM are trendy. These programs directly receive in the form of real numbers. Alike the binary version of IEAM, the program has three essential segments.

7.6.1 Defining Initial Population

IEAM (real-parameter version) is elementary program. Like IEAM (binary), we also generate the initial population in IEAM, but in this version, the solution can be directly developed in the form of real numbers that are also in the given bounds.

The first initial population is generated with the help of the following commands. First size of population and variables are declared:

```
ps = 10; and dim=3;
```

Then boundary values are created, and finally, the initial population is generated. Let us suppose that we want to develop all solutions in between –5 to 5.

```
xbound=5;
oldpop = 2 * xbound * rand(ps,DIM) - xbound*ones(ps,Dim);
```

Following the generation of solutions, fitness is evaluated in order to generate a new population and selection. The fitness values of the solution are evaluated using separate functions. This fitness assessment method is used to assess the fitness of the COCO BBOB framework.

```
fitness=feval(fun,oldpop');
```

7.6.2 Adaption Operator

For generating a new population, the adaption method is used. This is done by defining the adaption function for adaption operator.

```
function[new_pop,fitad]=adaption(oldpop,xbound,Dim,
fitness,ps,d1,c1)
xmin = -xbound * ones(1,Dim);
xmax = xbound * ones(1,Dim);
fitad=fitness;
favg1=mean(fitad);
favg=favg1*ones(ps,Dim);
fitad3=repmat(fitad,1,Dim);
c=fitad3./favg;
This is how c is generated, then
mb= oldpop(1,:)-oldpop(ps,:); (best -worst) is calculated.
Now for the first solution, the following equation is used
new_pop(1,:)=c(1,:).* oldpop(1,:)+rand(1,Dim)
This formula will update others.
for h=2:ps
new_pop(h,:)=oldpop(1,:)+rand(1,Dim).*mb
```

This is how adaptation is implemented.

Clamping of solutions:

Solutions generated during adaption may go out of bounds. The clamping process is used to send them in bounds.

Code for clamping:

```
s = new_pop < repmat(xmin,ps,1);
new_pop = (1-s).*new_pop + s.*repmat(xmin,ps,1);
b = new_pop > repmat(xmax,ps,1);
new_pop = (1-b).*new_pop + b.*repmat(xmax,ps,1);
```

7.6.3 Selection Operator

Alike the selection operator used in IEAM (binary), this selection also combines oldpop and new _ pop. It does a sort of the combined population and selects the best solutions.

Selection function with code:

```
function[sel_pop,fitsel]=selection(oldpop,new1_pop,ps,fun,Dim,
fitad1)
fin=cat(2,oldpop,fitad1);
x=feval(fun,new1_pop');
fin1=cat(2,new1_pop,x');
marge=cat(1,fin,fin1);
final_sort=sortrows(marge,Dim+1);
sel_pop=final_sort(1:ps,1:Dim+1);
fitsel=sel_pop(:,Dim+1);
sel_pop=sel_pop(1:ps,1:Dim);
end
```

Repeat the adaption and selection processes with the following code for many iterations:

```
for iter = 2 : maxiterations
[new_pop,fitad]=adaption(oldpop,xbound,Dim,fitness,ps,d1,c1);
new1_pop=new_pop;
new1_pop;
fitad1=fitad;
[sel_pop,fitsel]=selection(oldpop,new1_pop,ps,fun,Dim,fitad1);

oldpop=sel_pop;
fitness=fitsel;
if fbest > fitness(1)
   fbest = fitness(1);
   %xbest = par(1,:);
end
if feval(FUN, 'fbest') < ftarget
   break;
end
```

It can be seen from the code that in each iteration, the output of the adaption operator is passed to selection and the output of selection is passed to adaption.

So this is all about IEAM algorithms.

Future work: Although the real-parameter version of IEAM is very effective in solving optimization problems, the algorithm has several flaws. For starters, it performs poorly on high-dimensional functions. We need to figure out why this algorithm performs poorly in higher dimensions. What modifications must be made to this algorithm for it to be effective in solving higher dimension problems? Another issue with IEAM is that it is stagnant. It does not perform well on higher dimension problems due to a stagnation issue.

In the current version of EAM and IEAM, the solution is adapting to a familiar environment.

It was explained in the base paper of EAM, "A Bio Inspired Algorithm for Solving Optimization Problem," presented at the ICCCT conference and published by IEEE, MNNIT Allahabad India, that if changes in dynamic environment can be framed in the algorithm. It can find the best solution in a few iterations. That theory can also be used to improve the new version's effectiveness.

8 Other Important Optimization Algorithms

We reviewed numerous sorts of optimization problems and strategies for solving them in the previous chapter. We've shown how to make a nature-inspired algorithm. How can a natural-inspired technique be converted into an algorithm that can be used to address an optimization problem? How can a natural occurrence be translated into various mathematical functions? We've also talked about why nature-inspired algorithms are more efficient than other optimization techniques.

Although much has been said about nature-inspired algorithms, less has been said about how they may be tweaked to make them faster. Is it possible to build a new version of nature-inspired algorithms that incorporate previous optimization methods' beneficial features?

Scientists combined new/existing biological phenomena with favorable properties of other optimization algorithms to build a speedier version of the nature-inspired algorithm. So, to conduct effective research in nature-inspired algorithms, we must first discover how to merge the key qualities of existing methods into nature-inspired algorithms.

To grasp these qualities, we must examine alternative methodologies for building optimization algorithms and specifics of other nature-inspired algorithms not covered in this book.

This chapter offers an overview of current optimization strategies and a summary of important nature-inspired algorithms.

8.1 METHODS FOR CREATING AN OPTIMIZATION ALGORITHM

An optimization algorithm is used to identify the minimum/maximum value of the objective function by targeting the optimal global solution to the problem. To target this optimal global solution, some method has to be implemented in an algorithm. Two ways are widely used, that is, traditional methods and nature-inspired algorithms.

8.1.1 MATHEMATICAL METHODS

Many mathematical methods were used in the design of optimization algorithms prior to the creation of a nature-inspired algorithm. These mathematical techniques were used to direct the search toward the best overall solution. The only change made to the nature-inspired algorithm was to change the method from mathematical to biological. The rest of the steps are the same. For instance, for all optimization

DOI: 10.1201/9781003313649-8

algorithms, whether mathematical or natural, the first step is to divide the search space into a predetermined number of points. The second step is to evaluate selected solutions in order to guide the search. These solutions are evaluated using the mathematical or biological technique specified in the algorithm. The search direction is predicted based on these evaluations. Future mathematical plans can be divided into two categories: gradient-based methods and derivation-free methods.

8.1.1.1 Gradient-Based Methods

One popular method for solving optimization problems is to compute the function's derivative and set it to zero. The value at which the derivative becomes zero may produce the optimal solution value. The function's second derivative is calculated once more, and the optimal solution is identified based on the second derivative value. Because a computer cannot handle continuous values and only works with discrete input values, the derivative of a function on each input value is calculated using the gradient. In the gradient-based method, the gradient values of selected solutions are evaluated, and then directions toward a global optimal solution are identified. Generally, the algorithm designed for implementing a gradient-based approach is point-to-point-based. From a single solution, the search for the optimal global solution begins. A method for converting an existing value to a new value. This is accomplished by inserting a random value into the preceding value. The algorithm's added value is referred to as step size. Following the creation of a new solution, both solutions are compared, and a new search direction and step size are calculated. The method used to increase or decrease the step size and predict the search direction defines the algorithm type. For solving optimization problems, a variety of gradient-based methods are available. The steepest descent method, conjugate gradient method, nonlinear conjugate gradient method, Newton method, and quasi-Newton method are some algorithms. Although many variants use gradient-based techniques to guide the search, new derivation-free methods were developed because gradient-based methods are expensive.

8.1.1.2 Derivation-Free Method

Gradient-based methods evaluated the gradient value at selected points to identify the direction of the search. Other techniques that do not require the calculation of gradient are called as derivation-free methods. Some popular derivation-free methods are the interpolation method, the Hooke–Jeeves method, and pattern search methods. The process used in the derivation-free-based method is very similar to derivation-based methods. Like the gradient-based approach, derivation-free methods start their process by selecting some solutions from the entire search space. After choosing these solutions, they are evaluated to guide the search for the best solution position. The only change that has been made in the derivation-free method is in the evaluation process. In such methods, solutions are evaluated by calculating the value of the functions and the new value of the step size and directional variable is identified by the method (Hooke–Jeeves, pattern search, and others). Although these methods are not costly, they suffer from the problem of stagnation. They may stick in local optima. For removing this shortcoming, new heuristic-based methods have been implemented.

8.1.2 Search-Based Methods

8.1.2.1 Heuristic Search-Based Method

Heuristic-based methods address the shortcomings of existing mathematical techniques. Unlike other mathematical-based methods, they provide quick and efficient problem solutions. A heuristic function is used to guide the direction of search in a heuristic-based method. The heuristic function is usually dependent on the problem at hand, and it reduces the number of evaluations by discarding useless solutions. However, because the heuristic function is dependent on the problem. These methods cannot be generalized to solve other problems.

8.1.2.2 Metaheuristic Search-Based Method

Metaheuristic algorithms, like heuristic search methods, are used to solve optimization problems and find optimal solutions using heuristic functions. However, the heuristic function used in these algorithms is not problem-specific, as opposed to heuristic-based methods. The beauty of metaheuristic algorithms lies in their ability to design heuristic functions.

8.1.3 Special Search-Based Methods (Nature-Inspired Algorithms)

Nature-inspired algorithms belong to a specific category of metaheuristic algorithms in which a heuristic function is implemented by mapping some natural phenomenon. As a result, these algorithms are also known as nature-inspired algorithms. These algorithms are very popular these days due to their broad applicability and use. Another factor that motivates us to create nature-inspired algorithms is the complexity of the problems. Optimization problems are complex problems that necessitate a significant amount of computational effort and are prone to failure as the problem size grows. Nature provides efficient and effective solutions to many real-life problems. It can direct the search in the right direction and produce an optimal solution to the problem in a short period. Nature-inspired algorithms provide a very simple method for such complex problems, which can also be implemented in a very short time. These algorithms seek the global optimal solution to the problem while maintaining a balance between exploitation and exploration. The optimization problem is defined as finding the best possible/desirable solution. Many nature-inspired algorithms, such as genetic algorithms, particle swarm optimization, and ant colony optimization (ACO) algorithms, are simple to understand and used to solve various complex problems. Scientists are providing excellent solutions for cutting-edge applications due to the simplicity of these algorithms.

Many existing optimization algorithms were created by applying the biological theory of natural selection and evolution. Such algorithms are referred to as evolutionary algorithms. In addition to the evolutionary algorithm, other optimization algorithms were implemented by mapping the social intelligence. These algorithms are known as swarm intelligence–based algorithms. Earlier work on optimization algorithms focused on either evolutionary or swarm intelligence–based algorithms, but this is changing. Nowadays, many new optimization algorithms are being developed that do not belong to the evolutionary adaptation (EA) class or are based on swarm intelligence. Now people are mapping the characteristics of other domains like physics, sociology, and chemistry for designing optimization algorithm.

Optimization algorithms that are nature-inspired:
The advantage of designing nature-inspired algorithms follow:

- Each natural system defines how a set of solutions can be evolved to identify optimal solutions.
- The computation used in such a system is parallel and asynchronous.
- Their implementation requires a straightforward theory.
- The functionality of such a system is due to the collective efforts of their participants.

These properties help participants of systems work in parallel and independently for searching global information. Due to these properties, these natural systems are robust and can perform parallel computation independently[5–6]. These systems also provide freedoms to their participant to adapt according to the changing problem domain. Nowadays, people are also investigating how the environment affects the performance of such algorithms. How to apply adaptability in the algorithm?

Classification of nature-inspired algorithms:
Two categories are famous for the nature-inspired algorithm. The first classification of nature-inspired algorithms is done based on a biological method used for designing algorithms. The second classification is based on the number of solutions used to search for the optimal solution.

Classification based on biological methods:
Various scientists have developed many nature-inspired algorithms for solving optimization problems up to this point. These algorithms merely replicate the process used in real biological systems. Nature-inspired algorithms can be classified into six types based on the biological method used to design them. Nature-inspired approaches are classified into six major categories, which are as follows:

- Swarm Intelligence Based Algorithm
- Evolutionary algorithms
- Biological neural networks
- Physics-based algorithms
- Immune systems
- Cell-based algorithms

Classification based on the number of solutions used for searching for the optimal solution:
In addition to the previous classification, there is another that is widely used. This classification is based on the solutions required to start the randomized algorithm. They are divided into two types.

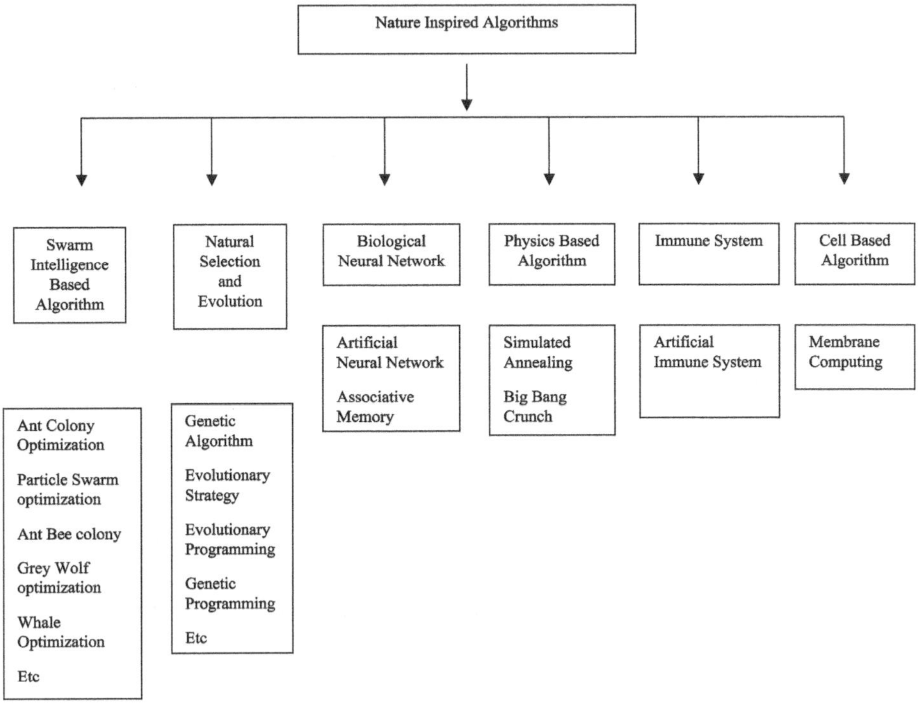

8.1.3.1 Point-to-Point-Based Algorithm

These algorithms begin by looking for a single solution. The search is conducted using the algorithm's defined operators, and a new solution is generated. This newly generated solution is compared to the previous one to determine whether it is converging correctly. If the new solution is acceptable, it is chosen; otherwise, a random number is generated to decide whether we should proceed to the next point. Simulation annealing and taboo search are two examples of point-to-point algorithms.

Let us look at some important point-to-point-based algorithms and how they can be used to solve optimization problems.

8.1.3.1.1 Simulated Annealing

The simulated annealing (SA) algorithm is a point-to-point algorithm. It is used to solve optimization problems. It begins its search with one solution and evolves it until the desired solution is not captured. The simulated annealing algorithm is based on the physical phenomenon of metal annealing. In physics, the annealing process is used to convert a molten metal into a crystal. The rate of crystal formation is affected by the rate of cooling. The goal of the annealing process is to obtain optimal structure by minimizing the energy of solid formation.

To convert a material into a crystal form, it is first heated to annealing temperature (the temperature at which the material begins to melt) and then cooled with a coolant until the crystal forms. To achieve an optimal structure with large crystals, the heated material should be cooled slowly. To reduce the energy of the crystal, slow cooling is required. The resulting structure will not have the minimum energy if the liquid is quickly cooled (quenched).

How it works:
Because the theory of the simulated annealing method is associated with energy minimization. This phenomenon can be used to solve problems involving depreciation. It is extremely helpful in resolving minimization problems. In a maximization problem, the objective function must first be converted into a minimization problem before the algorithm can be applied.

The algorithm begins with a random solution and selects two important parameters: temperature and number of iterations to optimize any function. At a specific temperature, this randomly generated solution is updated to generate a new solution, and this process can be repeated an infinite number of times.

Once a new solution has been generated, its fitness is compared to that of the previous solution. If the fitness of the new solution is less than the fitness of the previous solution, the new solution will be accepted. If the fitness of the new solution is greater than the fitness of the previous solution, the Boltzmann probability distribution Delta E is computed.

A random value is generated, and if it is less than the Boltzmann probability distribution, a new solution is accepted; otherwise, a new solution is generated.

Algorithm 8.1: Algorithm of an SA

1: **SA()**
2: **begin**
3: Initialize counter by assigning `count=1` and `count1 = 1`;
Choose rate of decreasing temperature `a`,
Choose initial temperature `T`,
number of iterations `n`,
number of repetitions `r`,
initial solution `x = rand(1,Dim)` // Initialize the particle's position with a random vector
4: Choose a new solution `x'` in the neighborhood of `x` by using some distribution
5: Evaluate the cost of `x` and `x'`
6: If fitness of `x'` is less than `x` goto step 9 else go to step 7
7: calculate Boltzman probability distribution and generate a random number
8: If (random number < Boltzman probability distribution) goto step 9 else goto 10
9: accept `x'` and replace `x` by `x'`
10: `Count=Count+1`
11: Repeat steps 4 to 10 until `Count ==n`
12: Replace `T` by `a*T`
13: `Count1=Count1 + 1`
14: Repeat 4 to 12 until `Count1==r`
25: **End**

8.1.3.1.2 Tabu Search

Tabu search is a straightforward metaheuristic search method for solving combinatorial optimization problems[14]. This algorithm begins with a random starting point. Let us denote this solution as x. Following the selection of the initial solution, all neighborhood solutions around x are generated. The fitness of these solutions is then compared to x, and a solution with a fitness greater than x is chosen as the next solution only if it is not on the tabu list. A tabu list is a list of recently visited solutions. To avoid convergent convergence on a local optimal solution, the newly generated best solution is checked against the tabu list. A tabu list is a collection of recently visited solutions. To avoid convergence on a local optimal solution, the newly generated best solution is checked in the tabu list. If it is on the tabu list, it will be rejected, and the best alternative will be chosen. This is done to avoid repeating searches for the same solutions and to avoid a stagnation problem. This process is repeated until the global optimal is captured or the termination condition is met. It is also possible that the fitness of the best solution chosen in the neighborhood of x is not better than x. In this case, we must generate a random number to determine whether we should proceed to the next solution or not. As a result, unlike other local search strategies, a bad move with low fitness can be accepted in some cases. We continue to explore the search space from one point x to another point x′, which is x's neighbor, ie x′N(x), where N is the neighbor function.

The stopping criteria for the tabu search can be a fixed number of iterations, a solution threshold, CPU time, a fixed number of iterations without any improvement, or iteration when no possible moves remain.

Tabu search has an advantage in which it allows non-improving solutions to be accepted in order to escape from the local optimum. Second, it can be applied to both discrete and noncontinuous search spaces. The only disadvantage with tabu search is that it can require a large number of iterations to find a global optimum.

Algorithm 8.2: Algorithm of a Tabu Search Algorithm

1: **TS()**
2: **begin**
3: Initialize counter by assigning `count=1`
initial solution `x = rand(1,Dim)`
and a `Tabu List={}` of size `k` elements
// Initialize solution with a random vector
4: Generate neighborhood solutions of `x` and define set of neighbors `n(x)`
5. Choose best solution x′ of set `n(x)` with minimum fitness
6: Evaluate the cost of x and x′
7: If fitness of x′ is less than `x` and x′ does not belong to `Tabu List` goto step 8 else goto 9
8: accept `x′,x=x′` and update `Tabu List`
9: `Count=Count+1`,update `n(x)` to `(n(x)-x′)`
10: Repeat steps 4 to 10 until `Count ==n`
11: **End**

8.1.3.2 Population-Based Algorithms

In addition to these point-to-point-based algorithms, population-based algorithms are also used for solving optimization problems. Population-based algorithms start their search with solutions and implement some natural phenomena to generate new solutions. Some population-based algorithms are designed by mapping the intelligence of swarms in a program. These algorithms are called swarm intelligence–based algorithms. Some popular swarm intelligence algorithms are particle swarm optimization and grey wolf optimization. We discuss the applications of these algorithms later in this book. Other swarm intelligence algorithms are ACO, ant–bee colony, whale optimization algorithms, bat optimization algorithms, cuckoo search algorithms, firefly algorithms, and ant–lion optimization algorithms. In this chapter, we give a brief overview of some of them.

Most of the early algorithms used for solving optimization algorithm comes under evolutionary algorithms. These algorithms implement evolutionary operators to search for the optimal solution. In general, three operators named selection, crossover, and mutation are used for solving the problem. The genetic algorithm (GA) was the most successful algorithm for solving optimization algorithms. We have covered some essential EAs, including GAs and differential evolution. Some other evolutionary algorithms are evolutionary programming (EP), evolutionary strategy (ES), and genetic programming (GP). Let us have a look on some popular evolutionary algorithms.

8.1.3.2.1 EAs

All algorithms that implement evolutionary learning are collectively referred to as EAs. EP, ES, genetic algorithms, and genetic programming algorithms all use separate concepts to implement evolutionary learning. All these algorithms can be used to solve optimization problems. When formulating an optimization problem solution, these approaches use different methods for evolving good solutions from a set of randomly generated solutions. Although their strategies may differ, their operators are nearly identical. All these techniques use a subset of selection crossover and mutation operators to identify an optimal solution.

Because of the properties of parallel computation and self-adaptation, EAs are ideal for solving optimization problems, particularly complicated problems that are deemed impractical to solve using traditional methods. Other benefits of using EAs include the lack of or limited need for information and the ease of implementing them.

8.1.3.2.1.1 ES

Rechenberg proposed the evolutionary strategy in 1960. It is a population-based algorithm, similar to a genetic algorithm. It, like other population-based algorithms, begins its search with a set of solutions. The initial population is defined by these solutions. The ES algorithm can be used to solve both discrete and continuous optimization problems, with and without constraints[18]. It excels at solving problems where defining a mathematical model is difficult.

To solve an optimization problem with an ES, first choose a representation for solutions. In an ES, a real-parameter representation of the solution is very popular.

In addition to solutions, some control parameters such as step size, the number of iterations, and the selection type are required to start the process. The parameters to be optimized are frequently represented by a vector of real numbers. Another vector of real numbers defines the strategy parameters, which control the mutation of the objective parameters.

In the classic version of ES, one parent produces one offspring in one generation. Offspring were created by using a mutation operator that remained constant across generations. The first version of this algorithm was known as two-membered ES. Later, several updated versions of this algorithm were proposed, all of which used a multi-membered approach. The simplest algorithm was the two-membered ES, which uses one parent to produce one offspring in one generation. Offspring were produced by using a mutation operator that remained constant across generations. In later versions, a set of parents is used to produce mutated offspring. A later version of this algorithm is also known as $(\lambda + \mu)$ ES, and its approach is very similar to the real-parameter GA.

Several versions of ES are popular in solving optimization problems. One version is known as $(\mu/\rho,\lambda)$-ES.

The first term specifies how many initial solutions will be chosen and how many parents will produce offspring. In $(\mu/\rho,\lambda)$-ES, μ solutions will be chosen at random, and ρ will take part in the production of offspring. Finally, λ offspring will be used to select parents for the following generation.

In ES, two operators' mutations and selections are used to produce offspring. During mutation, ρ solutions are mutated to produce λ offspring. Depending on the selection criteria used to select parents, two variants, $(\mu/\rho,\lambda)$-ES and $(\mu/\rho+,\lambda)$-ES, are popular. In the first variant, ρ best solutions are selected from λ offspring. In the second variant, the first parent population ρ is combined with λ offspring, and then the best ρ solutions are chosen on the basis of the fitness function.

In ES, the mutation operator is used to improve the algorithm's convergence rate. To create a mutated solution, choose a distribution and generate a new solution based on the step size. The size of the steps varies from generation to generation. Initially, the step size is set to be as large as possible to produce various solutions. This step size can be changed in later generations based on the needs of the problem.

Recently a new version of ES is proposed which is very popular. This new version of ES is called a covariance-matrix-adaptation-based ES (CMA-ES).

In this new version, the distribution around the mean solution is represented by a covariance matrix. This distribution is used to update the step size in mutation.

Algorithm 8.3: Algorithm of an ES

ES(μ/ρ,λ)
Begin
 `x = rand(μ,Dim)` // Initialize μ solutions
 evaluate the cost for each solution
Select best ρ for generating offspring
 Do

```
for ρ parent do
    Select λ offspring according to the values of control variable (step size, distribution)
End
cost _ x' = evaluate the function cost for all new solutions λ
Select ρ best solutions for nest generation
until stopping criterion is met
End
```

8.1.3.2.1.2 EP

In 1966, Lawrence J. Fogel proposed another evolutionary algorithm that is similar to ES and GAs. It operates on a similar principle. His son, David Fogel, expanded on this concept. The book *Artificial Intelligence through Simulated Evolution* is a seminal work in EP. The initial population in the evolutionary programming method is chosen at random. EP, like ES, generates offspring by performing the mutation operator. The fitness of each offspring solution is calculated. When the solution gets close to the global optimum, the strength of this mutation operator decreases.

It should be noted that the EP method does not use any crossover as a genetic operator. EP is used for the evolution of finite state machines. This algorithm follows the self-adaption technique to generate good solutions for given problems.

8.1.3.2.1.3 GP

Prof. John R. Koza created GP in 1992. GP is inspired by biological evolution to discover computer programs that perform user-defined tasks. The GP technique provides a framework for automatically generating a working computer program from a high-level problem statement. GP achieves automatic programming by genetically breeding a population of computer programs using Darwinian natural selection principles and biologically inspired operations. GP is a subset of GAs in which each individual is a computer programmer. GP is a machine learning technique used to optimize a population of computer programs based on a fitness landscape determined by a program's ability to perform a given computational task.

A typical genetic programming approach is composed of four steps:

1. Initial populating of the computer program's functions
2. Evaluating each program on the population and assigning it a fitness based on its accuracy in solving the problem
3. Developing new programs
    a. using crossover to create new programs from older ones
    b. by mutation to create new programs from old ones
    c. by directly copy the best programs from the current generation into the next generation
4. As a final result, choosing the program with the highest fitness value that has appeared so far in any generation

GAs are less common than GP. A GA produces a quantity that is either minimized or maximized, whereas GP produces a computer program. As a result, GP can be viewed as a program that generates another program.

GP is effective for a wide range of problems. It can find a near-optimal program that meets all the constraints. As an example, in a facial recognition system, we can find a program that is both fast and accurate. We can also use the GP techniques when we do not have adequate knowledge of the system that is to be solved.

8.1.3.2.2 Swarm Intelligence–Based Algorithms

As we have seen, many optimization algorithms have been developed using natural selection and evolution theory. This theory is not, however, the only theory that has been used to develop optimization algorithms. In addition to natural selection and evolution theory, swarm intelligence is used to develop optimization problems. Some trendy algorithms have been developed by mapping the intelligence of swarms. ACO, ant–bee colony, bat algorithms, particle swarm optimization, gray wolf optimization, and cuckoo search are swarm intelligence–based algorithms. In this section, we look at some of the most crucial swarm intelligence algorithms.

8.1.3.2.2.1 ACO

ACO algorithms are a popular swarm intelligence–based algorithm proposed by Marco Dorigo in his PhD thesis in 1992. The algorithm concept is based on the intelligent behavior displayed by a group of ants while searching for a food source. Foraging is a critical activity for any swarm. Swarms do this activity in groups because they can reduce their effort in finding a food source by communicating with one another. Communication is crucial in defining swarm intelligence. Several algorithms are developed by mapping the foraging activity of various swarms. Let us see how communication can be implemented in ant colony optimization.

Natural phenomena used for designing the algorithm:

An ACO algorithm is created by mapping the ants' foraging behavior. In ACO, it is assumed that ants live in colonies. They communicate with one another to search for a food source as a group. Initially, some ants from the colony choose a random path from the colony to the food source. Later, to reduce the effort required by the entire group to locate a food source, they share the information about their paths with other ants. These ants, which travel from colony to food source at random, communicate with other ants through the use of chemical trails known as pheromone trails. After reaching the food source, each ant returns to the colony via the same path, leaving a chemical pheromone trail in its wake. As time passes, the pheromone trail dissipates. A longer path will have less pheromone trail because more pheromone trail will evaporate as the ant spends more time returning to the colony. A shorter route will also have a stronger pheromone trail. A new ant that wants to visit a food source will check the intensity of pheromone trails on different paths and will most likely choose the path with the stronger pheromone trails. This cycle is repeated until each ant has reached the food source.

Mapping natural phenomena for designing the algorithm:

The mapping of ant foraging activity can be accomplished in three steps. In the first step, some random paths will be chosen to populate the system. These preliminary solutions can be represented using an appropriate encoding.

1. Initial Population: This is generated by defining an appropriate encoding for the problem. Because ACO is very useful in solving graph-related problems, permutation-based encoding is the best choice for these problems.
2. Choosing the best path for finding food: Each ant will analyze the intensity of pheromone deposited on different paths and probabilistically choose the best path. A combination of different edges can represent a path in graph problems. In such situations, at each point, the ant will choose the best edge probabilistically. This probabilistic way of path selection is given by Equation 8.1.

The probability of choosing an edge between node i to j will be as follows:

ACO path selection probability equation:

$$p_{i,j} = \frac{\left(\tau_{i,j}^{\alpha}\right)\left(\eta_{i,j}^{\beta}\right)}{\Sigma\left(\tau_{i,j}^{\alpha}\right)\left(\eta_{i,j}^{\beta}\right)}, \tag{8.1}$$

where

$\tau_{i,j}$ is the amount of pheromone on edge i, j,

α is a parameter to control the influence of $\tau_{i,j}$,

$\eta_{i,j}$ is the desirability of edge i, j (typically $1/d_{i,j}$), and

β is a parameter to control the influence of $\eta_{i,j}$.

3. Pheromone Update: This step is used to update the concentration of pheromone chemical and is given by Equation 8.2.

Pheromone update equation:

$$\tau_{i,j}^{k} = \left(1-\rho\right)\tau_{i,j} + \ \tau_{i,j}, \tag{8.2}$$

where

$\tau_{i,j}$ is the amount of pheromone on a given edge i,j;

ρ is the rate of pheromone evaporation; and

$\tau_{i,j}$ is the amount of pheromone deposited, typically given by

$$\tau_{i,j}^{k} = \begin{cases} 1/L_k, \textit{if ant } k \textit{ travels on the edge } i,j \\ 0, \qquad \textit{Otherwise} \end{cases}, \tag{8.3}$$

where L_k is the cost of the kth ant's tour.

The three highlighted steps are repeated until the optimization problem has converged or otherwise terminated via a prespecified termination condition.

ACO was the first algorithm that aimed to search for an optimal path in a graph based on the behavior of ants seeking a path between their colony and a source of food. ACO can't perform well in large search space and is criticized due to its slow performance.

8.1.3.2.2.2 Artificial Bee Colony

The artificial bee colony algorithm (ABC) is also a swarm intelligence–based algorithm; it implements bees' foraging behavior for solving an optimization problem. It was proposed by Davis Karaboga in 2005. In the ABC algorithm, bees are categorized into three categories according to their search behavior when searching for food sources. These three groups of bees are

a. employed bees,
b. onlookers, and
c. scouts.

Each bee will be either an unemployed bee or an employed bee.

A bee is called an unemployed bee if it is searching for a food source. A bee who has acquired a food source is called an employed bee. An employed bee is always looking for new food sources near previously visited food sources. An unemployed bee may be a scout or an onlooker, depending on her strategy to find a food source. Scout bees are unemployed bees that begin searching for food in any random direction. An onlooker bee is a bee that searches for food with the assistance of another bee. An onlooker bee observes the employed bee's dance to determine which source it should exploit. The duration of the employed bee's dance informs the onlooker bee about the amount of nectar in the food source. The more time a bee is employed, the more nectar will be available as a food source. The employed bee becomes a scout if its food source has been abandoned. The onlooker bees use the nectar information shared by employed bees to choose a food source according to its nectar (fitness) amount. They select one of the food sources whose corresponding nectar is the highest.

Employed bees will exploit these food sources, and new solutions will be generated. A greedy selection will be used to replace an existing food source with a new food source if the fitness of the new solution is superior to the previous solution. After generating new food sources, employed bees enter the dance floor and begin dancing. The onlooker bee will create a new food source near an existing food source based on the quality of nectar associated with each employed bee. Again, a greedy selection is required to replace an existing food source with a new one if the fitness of the new solution is good. Again, a greedy selection is required to replace an existing food source with a new one if the fitness of the new solution is good. While updating the solution, a counter is attached to each solution that counts the total number of failures and assigns a fixed value to it. This counter is updated whenever the fitness of a new solution is inferior to that of a previous solution. In this case, the solution will remain unchanged. This process is repeated until the counter reaches its maximum value.

If the counter reaches its maximum value, the scout bee phase is triggered. During the scout phase, one solution will be replaced by a random solution.

Algorithm 8.4: Algorithm of an ABC()

```
   Function ABC()
   {
   count=0; % Initialize counter
1. Create the initial population by generating n number of solutions in given bounds
   (food sources)
2. Apply the employed bee phase and generate new solutions around already previ-
   ous solution
3. Apply greedy selection
   for i=1 to n do
   if(fitness of new solution)>(fitness of previous solution)
   solution = new solution;
   Else
   solution = old solution;
   count=count+1;
4. Calculate nectar quality against each solution by measuring fi/Σfi
5. Apply onlooker phase
     Repeat (while fi > fi/Σfi) skip
     else select the solution
6. Generate a new solution around selected solution
   if(fitness of new solution)>(fitness of previous solution)
   solution = new solution;
   Else
   soultion=old solution;
   count=count+1;
   End
   if (count > fixed value)
7. Apply scout phase
8. Generate a random solution and replace previous solution
   End
```

In this way, the ACO finds the global best solution.

ACO has various advantages like it has a guaranteed convergence, and it is relatively fast as compared to other algorithms of its class.

8.1.3.2.2.3 Bat Optimization Algorithms

After partial swarm optimization, several swarm intelligence algorithms were designed by mapping the foraging behavior of other swarms; the bat algorithm is one of them. It implements the intelligence behavior of bats for solving an optimization problem.

The bat algorithm was proposed by Xhi-She in 2010. It is inspired by the echolocation behavior of microbats with a varying pulse rate of emission and loudness. Bats use sonar waves to identify the location of prey. While searching the prey, they produce sound and emit pulse rate. Sound and pulse rate vary according to the distance of the prey.

They emit sonar waves to determine the location of prey. After hitting the objects, these waves are reflected back to the bat. It is assumed that bats can distinguish between reflected rays from prey and reflected rays from obstacles. They adjust their positions by examining the characteristics of the reflected wave. The formula for defining a new position is shown in Equations 8.4, 8.5, and 8.6:

$$F_i = f_{min} + \left(f_{max} - f_{min}\right) \times \beta \tag{8.4}$$

$$V_i(t) = V_i(t-1) + \left(x_i(t-1) - x*\right) * f_i \tag{8.5}$$

$$x_i(t) = x_i(t-1) + v_i(t) \tag{8.6}$$

8.1.3.2.2.4 Firefly Optimization Algorithms

Professor Xin-She Yang developed a firefly optimization algorithm by mapping the light-flashing behavior of fireflies. Fireflies are insects found in tropical environments, and about 2000 species of this insect are found on earth. They have wings and can produce light chemically. This light may be in the form of green, yellow, or pale red in color. They flash light either to attract a partner or prey or sometimes as a warning signal.

FIGURE 8.1 Fireflies on a full moon night.

The position of the firefly is represented by a solution vector in the firefly optimization algorithm. The algorithm's initial population is created by randomly selecting firefly positions within the given bounds. Each firefly is drawn to the brighter firefly. This attraction causes changes in the position of the firefly, and this attraction updates its position to get a new position close to a brighter firefly. This attraction weakens as the distance between fireflies grows. When there are no more luminous fireflies available, it may move to a random position. The position shift occurs either to search for a random position or to approach a better solution. The light intensity of a firefly can be calculated by examining the fitness of each firefly. The position of the solution determines it. It decreases as one travels farther away. So, if I_0 is the intensity of light at zero distance, then the intensity at d distance can be calculated using the following formula:

$$I_d = I_0 / d^2. \tag{8.7}$$

If absorption coefficient λ is also taken into account, then the formula will be represented as

$$I_d = I_0 \times e^{-\lambda d^2}. \tag{8.8}$$

The position of firefly X_i is updated near to a brighter firefly X_j using this formula:

$$x_i^{t+1} = x_i^t + \beta_0 e^{-\gamma r_{ij}^2} \left(x_j^t - x_i^t\right) + \alpha_t \epsilon_i^t, \tag{8.9}$$

where α_t is a randomization parameter and ϵ_i^t is chosen by Gaussian distribution.

If no brighter fly exists, then the random movement will be defined by

$$x_i^{t+1} = x_i^t + \alpha_t \epsilon_i^t. \tag{8.10}$$

The fitness of the new position is calculated, and the better one is moved in the next iteration.

8.1.3.2.2.5 Cuckoo Search Algorithms

In 2009, Xin-She Yang and Subhas Dev created the cuckoo search optimization algorithm. It was created by mapping the egg-laying strategy of brood parasite cuckoo birds. Some cuckoo birds are holoparasites, meaning their life cycle is entirely dependent on other host birds. They even lay their eggs in the nests of host birds to increase their population. In order to place their egg in a host nest, they will search for the best host nest whose egg resembles the shape of their egg. By removing some host eggs, they place their eggs in the best matching host nest. They lay their eggs in a host nest in such a way that their eggs hatch before the host eggs. These properties increase the chances of their survival and reduce the risk of damage.

Sometimes the cuckoo egg can be discovered by the host bird. In that case, either the egg is thrown out from the nest or the nest is abandoned and a new one made.

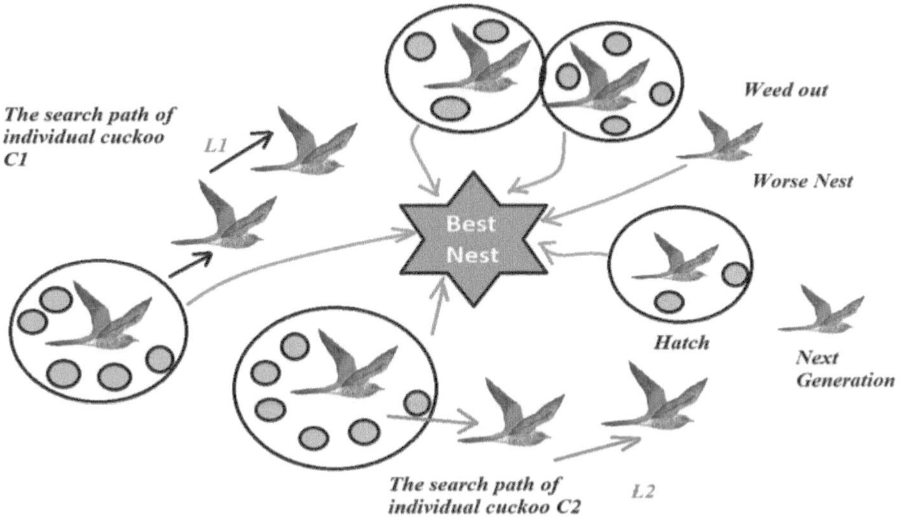

FIGURE 8.2 Cuckoo egg-laying strategy.

The cuckoo search program is implemented by beginning with some random host nests. These nests will provide the algorithm with its initial population. As previously stated, cuckoo birds will lay their eggs in the best nest. We will calculate the fitness of each solution to determine the best nest. The cuckoo bird moves from one nest to another until she finds the best nest. Because the movement of the cuckoo bird is random. It was incorporated into the algorithm through the use of levy flight. The step size of these random walks will be determined using Levy's distribution.

These three steps summarize the egg-laying strategy:

1. Each cuckoo bird searches for a suitable nest and places an egg in it.
2. The best nests, with high-quality nests, will be transferred to next generation. The quality of each nest will be determined on the basis of its fitness value.
3. The host bird can sometimes discover the cuckoo egg. Pa represents the probability of finding a cuckoo egg. If the host finds the egg, it can throw the egg out or abandon the nest and create a new one.

An algorithm can be summarized in the following steps.

Algorithm 8.5: Algorithm of a Cuckoo Search

---

1. Generate the initial population by randomly creating n host nests. Each host nest contains one egg. The quality of each host nest can be checked by calculating the fitness function. $f(x), x = (x_1, \ldots, x_d)^T$;

   In the first step, a cuckoo bird generates a new egg around the host nest for each host nest.

2. The position of this egg will be determined by the following equation:

   If $X_i(t)$ is the position of the host egg, a cuckoo egg will be generated at the position $X_i(t+1)$.

   $X_i(t+1)=X_i(t)+C_i(t)$,

   where $C_i(t)=0.01\,S_i(t)\oplus\left(X_i(t)-X_{best}(t)\right)$,

   where $S_i(t)=\frac{u}{|v|^{1/\beta}}$,

   where u and v are n-dimensional vectors and β is 3/2,
   and the elements u and v can be calculated by following normal distributions:

   $u \sim N\left(0,\sigma_u^2\right)$ and $v \sim N\left(0,\sigma_v^2\right)$

   $$\sigma_u = \left(\frac{\Gamma(1+\beta)\,.\,\sin\left(\pi * \beta/2\right)}{\Gamma((1+\beta)/2)\,.\,\beta\,.2^{(\beta-1)/2}} \right)^{1/\beta} \text{ and } \sigma_v = 1.$$

3. Egg replacement:
   The equation of egg replacement is determined by checking a random value if r is less than Pa, then the same egg is selected
   Else a new position is determined by

   $X_i^{t+1} = X_i^t + rand * \left(X_{d1}^t - X_{d2}^t\right).$

8.1.3.2.2.6 Whale Optimization Algorithms

The whale optimization algorithm is one of several swarm intelligence–based algorithms proposed by Saiyed Ali Mirjalili. The whale optimization algorithm mimics the intelligence used by whales when attacking prey.

Inspiration:
This algorithm was created by mapping the hunting behavior of whales. Before we get into the algorithm, let's talk about whales and how they hunt. There are numerous similarities between whales and humans. Whales, for example, have normal cells in certain areas of their brains that are similar to those found in humans. These cells are known as spindle cells. These cells are in charge of decision-making, emotions, and social behavior. In other words, our spinning cells distinguish us from other creatures. Whales have twice as many of these cells as an adult, which contributes significantly to their ingenuity. Whales have been shown to think, learn, judge, communicate, and experience human-like emotions, albeit at a deficient level of intelligence. One of the world's largest baleen whales is the humpback whale. The size of a humpback whale is comparable to that of a school bus. Its favorite foods include krill and shellfish. Humpback whales are fascinating because of their distinct hunting style. This type of eating behavior is referred to as the bubble-net search method.

FIGURE 8.3 Whale hunting strategy.

Hunting Strategy Followed by Humpback Whales:
Whales hunt their prey using a unique mechanism. The whales' hunting method is known as the bubble-net search method. The bubble-net search method is divided into two steps: upward spirals and downward loops. The first step is for whales to dive about 12 meters and create bubbles in a spiral shape around the prey before swimming up toward the surface. The second step is divided into three sections: coral loop, lobtail, and capture loop.

Mathematical modeling of the whale hunting process: Whale hunting can be modeled in two steps: enrichment and the bubble net method.

Enriching: The enriching process is used by humpback whales to identify the location of prey and enrich them. During enriching, it is assumed that prey is either at the best position or somewhere close to the best position. Whales update their positions according to Equations 8.11 and 8.12.

The following equations are used for enriching prey.

$$\vec{D} = \left|\vec{D}.\vec{X}_p(t) - \vec{X}(t)\right| \tag{8.11}$$

$$\vec{X}(t+1) = \left|\vec{X}_p(t) - \vec{A}.\vec{D}\right|, \tag{8.12}$$

where t indicates the current generation, $\vec{A}$ and $\vec{C}$ are coefficient vectors, $\vec{X}_p$ is the position vector of prey, and $\vec{X}$ states the position vector of a GWO.

Hunting Process: Two processes are used to carry out the bubble-net attacking method. The first process is known as shrinking the prey, and the second process is known as spiral updating position. Humpback whales swim in a shrinking circle around their prey while moving in a spiral pattern. A probabilistic equation is used to model this simultaneous behavior. During hunting, the whale updates its position using either the shrinking encircling mechanism or the spiral model. The new position of humpback whales can be represented by Equation 8.13, where p is a random number between [0,1].

When $p < 0.5$,

$$X(t+1) = \vec{X}*(t) - \vec{A} \cdot \vec{D}; \tag{8.13}$$

else

$$X(t+1) = \vec{D'} \cdot e^{bl} \cdot \cos(2\pi l) + \vec{X}*(t). \tag{8.14}$$

8.1.3.2.2.7 Harris Hawk Optimization Algorithms

The Harris hawk optimization algorithm is developed by Ali Asghar Heidari and coauthors for solving optimization problems. This algorithm is implemented by mapping the cooperative behavior and the chasing style of Harris hawks. The Harris hawk is an intelligent bird that hunts in a group. While hunting, they communicate with each other and attack prey from different directions. Moreover, they can change their chasing pattern according to the escaping strategy followed by the target.

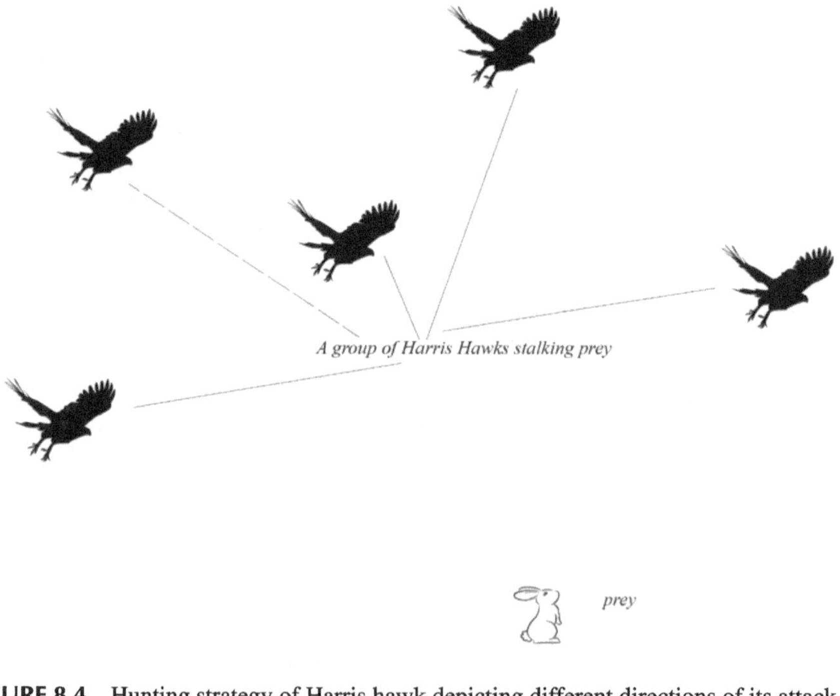

FIGURE 8.4 Hunting strategy of Harris hawk depicting different directions of its attack.

8.1.3.2.3 Other Important Optimization Algorithms

8.1.3.2.3.1 Teaching–Learning-Based Optimization Algorithms

In 2011, Rao proposed the teaching–learning-based optimization (TLBO) algorithm. The TLBO algorithm's concept is based on the teaching-learning process, which is used to increase students' knowledge.

Students can improve their knowledge, according to Rao, by using two methods of learning. They can learn from a teacher using the first method. This learning phase was modeled after the teacher phase, according to him. Learning by interacting with other students is another method used for learning. The learner phase is referred to as such.

TLBO, like other population-based algorithms, starts its search with some randomly generated solutions. These solutions comprise the initial population of learners. Several classroom sessions are used to refresh these students' knowledge. As a result, the learning process is divided into several steps, which are referred to as iterations in algorithms. The teaching and learning process is used to improve the learners' knowledge in each iteration. This process is repeated until the optimal solution is found. In each iteration, teaching and learning are implemented in two phases: (1) the teacher phase and (2) the learner phase.

During these phases, the population is divided into two classes: teachers and learners.

This division is made based on the fitness of each solution; the best solution in the population is chosen as a teacher, while the other solutions are chosen as learners.

Teacher Phase: The goal of the teacher phase is to update the mean knowledge of the entire class.

During this phase, some random solutions are initialized to form a group of learners in order to design optimization algorithms. This group of learners serves as the algorithm's initial population. These solutions improve their values by involving them in the teaching and learning process.

The entire population is divided into two categories for the purpose of generating new populations. This division is based on the fitness of the solutions. The teacher is thought to be the best solution in the entire population. Other solutions are regarded as students.

The learning of students is implemented in two phases: the teacher phase and the learner phase.

1. **Teacher phase**
   Students learn from the teacher during the teacher phase. The ultimate goal of the teacher phase is to increase the class's average knowledge. The change in knowledge is calculated by looking at the change in the mean result of the class in the subject taught by them based on their capability. Assume that there are 'm' subjects (i.e., design variables) and 'n' learners (i.e., population size, `k=1,2, . . . , n`) in any iteration 'I', and $M_{j,i}$ is the mean result of the learners in a specific subject 'j' `(j=1,2, . . . , m)`. The best overall result $X_{total} - k_{best,i}$ is the result of best learner k_{best} when all subjects obtained in the entire population of learners are considered together. However, because the

teacher is typically thought of as a highly learned person who trains learners to achieve better results, the algorithm considers the best learner identified as the teacher. The difference between the existing mean result of each subject and the corresponding result of the teacher for each subject is given by

$$Difference\_Mean_{j,k,i} = r_i\left(X_j, k_{best,i} - TFM_{j,i}\right), \tag{8.15}$$

where $X_j, k_{best,i}$ is the result of the best learner in subject `j`. *TF* is the teaching factor that decides the value of mean to be changed, and r_i is the random number in the range `[0, 1]`. Value of *TF* can be either `1` or `2`. The value of *TF* is decided randomly with equal probability as

$$TF = round\left[1 + rand(0,1)\{2-1\}\right]. \tag{8.16}$$

TF is not a parameter of the TLBO algorithm. The value of *TF* is not given as an input to the algorithm and its value is randomly decided by the algorithm using Equation 8.16. After conducting a number of experiments on many benchmark functions, it is concluded that the algorithm performs better if the value of *TF* is between `1` and `2`. However, the algorithm is found to perform much better if the value of *TF* is either `1` or `2` and hence to simplify the algorithm, the teaching factor is suggested to take either `1` or `2` depending on the rounding up criteria given by Equation 8.16. According to the $Difference\_Mean_{j,k,i}$, the existing solution is updated in the teacher phase based on the following expression:

$$X'_{j,k,i} = X_{j,k,i} + Difference\_Mean_{j,k,i}, \tag{8.17}$$

where $X'_{j,k,i}$ is the updated value of $X_{j,k,i}$. $X'_{j,k,i}$ is accepted if it gives a better function value. All the accepted function values at the end of the teacher phase are maintained, and these values become the input to the learner phase. The learner phase depends on the teacher phase.

2. Learner Phase

It is the second part of the algorithm where learners increase their knowledge by interacting among themselves. A learner interacts randomly with other learners to enhance its knowledge. A learner learns new things if the different learner has more ability than they do. Considering a population size of 'n', the learning phenomenon of this phase is explained in the following.

Randomly select two learners, `P` and `Q`, such that $X'_{total} - P_i \neq X'_{total} - Q_i$ (where $X'_{total} - P_i$ and $X'_{total} - Q_i$ are the updated function values of $X_{total} - P_i$ and $X_{total} - Q_i$ of *P* and *Q*, respectively, at the end of the teacher phase).

$$X''_j P_i = X'_j, P_i + r_i\left(X'_j, P_i - X'_j, Q_i\right), \quad if\ X'_{total} - P_i < X'_{total} - Q \tag{8.18}$$

$$X_j''P_i = X_j',P_i + r_i\left(X_j',Q_i - X_j',P_i\right), \quad if\ X'_{total} - Q_i < X'_{total} - P_i \tag{8.19}$$

$X_j''P_i$ is accepted if it gives a better function value.

Equations 8.18 and 8.19 are for minimization problems. In the case of maximization problems, Equations 8.20 and 8.21 are used.

$$X_j''P_i = X_j',P_i + r_i\left(X_j',P_i - X_j',Q_i\right), \quad if\ X'_{total} - Q_i < X'_{total} - P_i \tag{8.20}$$

$$X_j''P_i = X_j',P_i + r_i\left(X_j',P_i - X_j',Q_i\right), \quad if\ X'_{total} - P_i < X'_{total} - Q_i, \tag{8.21}$$

which does not necessitate any algorithm-specific parameters for its operation. The TLBO algorithm requires only common controlling parameters such as population size and generation number. The TLBO algorithm has gained widespread acceptance among optimization researchers.

The TLBO algorithm is a population-based algorithm that simulates the classroom teaching–learning process. This algorithm requires only the common control parameters, such as population size and the number of generations, and does not require any algorithm-specific control parameters.

8.1.3.2.3.2 Big Bang–Big Crunch Algorithms

Big Bang–Big Crunch (BB-BC) algorithms are based on the theory of the Big Bang Crunch, which is used to explain the theory of evolution of the universe. It also explains how the birth of the universe has taken place. It also explains what will be the future of our universe and how it will end. The two main phases of this theory are the Big Bang and the Big Crunch. The Big Bang phase consists of procedures

FIGURE 8.5 The big bang phase where probable solutions are spread throughout the search space.

of energy dissipation in the nature term of disordering and randomness. The Big Crunch phase is a procedure that randomly distributes particles and draws them into an order.

These two steps are implemented by the BB-BC algorithm when solving optimization problems.

Each iteration includes the use of two operators, big bang and big crunch. The Big Bang phase algorithm generates random solutions, which are then combined to form a single envoy point in the Big Crunch phase. The value of this single point is calculated by calculating the population center of mass.

At the start of the Big Crunch phase, the center of mass can be written as

$$\vec{x}^c = \frac{\sum_{i=1}^{N} \frac{1}{f^i} \vec{x}^i}{\sum_{i=1}^{N} \frac{1}{f^i}}. \tag{8.22}$$

The detailed algorithm follows.

Algorithm 8.6: Algorithm of a BB-BC Optimization

`Function BB-BC(){`

1: Big Bang Phase {

Generate `n` solution between the given bounds `[a,b]`

Calculate the fitness function values of all the solutions.

}

2: Big crunch phase {

Find the center of mass according to Equation 8.22.

3: The best-fit individual can be chosen as the center of mass instead of using Equation 8.22.

4: Calculate new candidates around the center of mass by adding or subtracting a normal random number whose value decreases as the iterations elapse,

where xc stands for center of mass, l is the upper limit of the parameter, r is a normal random number, and *k* is the iteration step. Then new point x_{new} is upper- and lower-bounded.

5: Return to step 2 until stopping criteria have been met.

}

8.1.3.2.3.3 Harmony Search

Geem in 2001 designed the harmony search algorithm. It maps the music improvisation process used by musicians for creating a pleasant harmony for solving optimization problems.

FIGURE 8.6 Music notes on a book.

During a music improvisation session, a group of musicians attempts to tune their instruments' pitch to create a pleasing harmony. This pleasant harmony produced by optimal pitch tuning of various instruments is analogous to the optimal solution in an optimization problem. The process of harmony search begins with n randomly generated solutions that represent a set of random harmonies. Each harmony is composed of a tuned pitch of d number of instruments, where d is the dimension of each solution. The pitch range of any instrument defines the variable's input domain. A musical instrument serves as the decision variable, harmony serves as the solution vector, and practice corresponds to iterations. The following is a more detailed description of the harmony search. A group of musicians collaborates to create a pleasing harmony. However, creating pleasant harmony is a difficult task. It takes a great deal of practice. In each practice, musicians will either try a pitch that is close to an existing pitch or generate it at random. They will keep some harmonies in their harmony memory to use with an existing pitch. To avoid the problem of pitch stagnation, pitch values are generated around previously stored pitches with some linear adjustments according to the equation:

$$x_{new} = x_{old} + b_{range} \times \varepsilon, \tag{8.23}$$

where

x_{old} is the existing pitch or solution from the harmony memory and

x_{new} is the new pitch after the pitch adjusting the action.

The overall working of a harmony search can be explained by the following pseudocode.

Algorithm 8.7: Algorithm of a Harmony Search

---

1. Initialization steps of Harmony Search
   a. **for** (i=1; $i \leq$ HMS; i++)
   i. **for** (j= 1; $j \leq n$; j++)
   1. Randomly initialize x_j^i in HM.
   ii. **endfor**
   b. endfor
2. Repeat below until a preset termination criterion is met.
3. Construction and evaluation of new solution candidate x
   for (j= 1; $j \leq n$; j++)
   if (rand(0, 1) < HMCR)
   Let x_j in x be the jth dimension of a randomly selected HM member.
   if (rand(0, 1) < PAR)
   Apply pitch adjustment distance bw to mutate x_j :
   $x_j = x_j \pm$ rand(0, 1) × bw.
   endif
   Else
   Let x_j in x be a random value.
   endif
   Endfor
   Evaluate the fitness of x: f(x).

4. HM update
   if (f(x) is better than the fitness of the worst HM member)
   Replace the worst HM member with x.
   Else
   Disregard x.
   Endif

8.1.3.2.3.4 Gravitational Search Algorithm The gravitational search algorithm (GSA) was proposed by Esmat Rashedi and coauthors in 2009. This algorithm was designed by mapping the law of gravity and mass interaction.

Law of gravity: This law is a very popular law in physics and was given by Newton. It states that each particle attracts every other particle with a gravitational force.

The gravitational force *F* between two masses M_a and M_b is directly proportional to the product of the masses and inversely proportional to the square of distance *R* between them. So *F* can be written as

$$F = G * M_a * M_b / R^2. \tag{8.24}$$

Also, according to other newton law, when a force is applied on a mass, acceleration takes place.

So, the acceleration of mass M due to gravitation force will be

$$a = F / M. \tag{8.25}$$

Also, the gravitation constant decreases with an increase in the age of universe, so

$$G(t) = G(t_0) * (t / t_0)^{\beta}. \tag{8.26}$$

Also, three kinds of masses are defined in theoretical physics:

1. Active mass: M_a is the measure of the strength of gravitational field due to a particular object.
2. Passive mass: M_p is the measure of the strength of an object's interaction with the gravitational field.
3. Inertia mass: M_i is a measure of resistance to the current state when force is applied.

Although in most cases we assume that all masses are equal; that is, $M_a = M_p = M_i$. If force between two masses, M_a and M_p, is represented by the following equation:

$$F_{ij} = G \frac{M_{aj} \times M_{pi}}{R^2}. \tag{8.27}$$

Then,

$$a_i = F_{ij} / M_{ii}. \tag{8.28}$$

For implementing GSA, the equation of force has been modified and R^2 was replaced by *R*.

It was due to experiments that were giving good results on *R*.

Now it is clear that two masses attract each other by gravitational force. Due to this force, both masses will accelerate toward each other. However, velocity changes in the heavier mass will be less while the light object will feel more changes in the velocity. In GSA algorithms, the position of mass is represented by the position of the solution, and the gravitational and inertial masses are represented by the fitness.

The following steps explain the algorithm:

Algorithm 8.8: Algorithm of a GSA()

Create the initial population of n position of masses by randomly generating a dth dimensional vector for each position. The position of ith mass can be defined by

$$X_i = \left(x_i^1, \ldots, x_i^d, \ldots\ldots, x_i^n\right), \tag{8.29}$$

where x_i^d represents the position of ith agent in the dth dimension.

---

1. At a time t, the force acting along dth dimension on mass m_i due to mass m_j will be represented by following equation:

$$F_{ij}^d(t) = G(t)\frac{M_{pi}(t) \times M_{aj}(t)}{R_{ij}(t) + \varepsilon}\left(x_j^d(t) - x_i^d(t)\right), \tag{8.30}$$

where $R_{ij}(t)$ is the Euclidian distance between two masses and ε is the small constant.

2. Total forces exerted on mass i can be written as

$$F_i^d(t) = \sum_{j=1,\, j\neq i}^{N} rand_j F_{ij}^d(t). \tag{8.31}$$

3. Hence, by the law of motion,

$$a_i^d(t) = {F_i^d(t)}\Big/{M_{ii}(t)}. \tag{8.32}$$

4. Due to this, the next position of the object will be

$$v_i^d(t+1) = rand_i \times v_i^d(t) + a_i^d(t) \tag{8.33}$$

$$x_i^d(t+1) = x_i^d(t) + v_i^d(t+1), \tag{8.34}$$

where $rand(i)$ is a uniform random variable in between $[0,1]$.

8.1.3.2.3.5 Black Hole Optimization Algorithm

Abdolreza Hatamlou proposed the black hole optimization algorithm in 2013 to solve optimization problems. It is based on the black hole theory. A black hole is a massive black star that generates an extremely powerful magnetic field around itself. Because of this strong magnetic field, other stars are drawn to black holes and shallowed if they come into contact with them. As the initial population of stars for the optimization algorithm, a set of random solutions will be used. To determine the location of the black hole, the fitness of each solution is calculated, and the best solution is selected as the black hole. Because of this black hole, all other stars become attracted to it and move closer to it.

Change in position experienced by a star X_i is shown by following Equation 8.35:

FIGURE 8.7 A huge blackhole at the center with star remnants revolving it at high speeds far from the event horizon.

$$X_i(t+1) = X_i(t) + rand * (X_{bh} - X_i(t)). \tag{8.35}$$

If in a new iteration, if the new position of any star becomes better than bh then that star will be a black hole. In addition, sometimes stars may cross the event horizon and be sucked by the black hole.

The radius of the event horizon is calculated by R as in Equation 8.36:

$$R = fh / \sum fi. \tag{8.36}$$

When the distance between the black hole and the other star becomes less than R, the candidate is collapsed, and a new star is created at a random position.

Algorithm 8.9: Algorithm of a Black Hole Optimization ()

Initialize a population of stars with random locations in the search space.

Loop

For each star, evaluate the objective function.

Select the best star that has the best fitness value as the black hole.

Change the location of each star according to Equation 8.35.

If a star reaches a location with lower cost than the black hole, exchange their locations/

If a star crosses the event horizon of the black hole, replace it with a new star in a random location in the search space.

If a termination criterion (a maximum number of iterations or a sufficiently good fitness) is met,

End loop

---

9 Application of Genetic Algorithms, Partial Swarm Optimization, and Differential Evolution in Software Testing

9.1 INTRODUCTION

In the book's first part, we have already discussed several optimization algorithms, such as the genetic algorithm (GA), differential evolution (DE) algorithm, particle swarm optimization (PSO), grey wolf optimization, and environmental adaptation method. After learning these optimization algorithms, you may wonder why we need to know so many optimization algorithms. Since all algorithms are used to solve optimization problems, why aren't we focusing on a single algorithm that works well for all optimization problems? Unfortunately, according to the well-known optimization theorem known as the no-free-lunch theorem, no optimization algorithm is superior in solving all optimization problems. For some problems, one algorithm may suffice. Others may perform well in other situations. Another explanation for not having a single optimization algorithm for solving all optimization problems is related to the characteristics of optimization problems. For example, in some optimization problems, there will be only one global optimal solution. These problems are referred to as unimodal problems. In other optimization problems, there may be multiple global or local optimal solutions. Due to the differences in characteristics of optimization problems, no single algorithm can be chosen as the best solution to all these problems.

Several optimization algorithms have been designed in the literature to solve real-world problems. These algorithms are generalized and can be used to solve any optimization problem. Their performance, however, will be determined by the design of the operators and the type of optimization problem. An algorithm whose operators exploit the promising regions is a good candidate for solving unimodal problems. These algorithms are designed by mapping natural phenomena with the assistance of operators, with the dominant operator focusing on exploitation. Other algorithms, in which dominating operators are more focused on searching the entire search space

DOI: 10.1201/9781003313649-9

to identify optimal solutions, will be suitable for handling multimodal problems. There is no predefined method to balance the process of exploitation and exploration. Therefore, these algorithms perform differently on different issues.

Inspired by the no-free-lunch theorem, scientists are creating new optimization algorithms in which they are trying to balance both exploitation and exploration. Such an algorithm may perform well in both kinds of problems, that is, unimodal and multimodal problems, and can cover most optimization problems.

In addition to creating new algorithms, people are also interested in using these algorithms for solving real-life problems. Take an example of a software engineer. A software engineer always wants to write an algorithm that can solve a real-life optimization problem. In contrast, those who are not software engineers want to learn how they can use these optimization algorithms to solve their optimization problems.

Now for learning how can we use these algorithms for solving optimization problems? Let us discuss some real-life applications of existing optimization algorithms. First, we discuss the problem and then show why that problem will be an optimization problem. After that, we discuss how an optimization algorithm can be used in solving the problem.

Several optimization problems from various engineering fields can be solved using these optimization algorithms. Many such optimization problems fall under the purview of chemical engineering, mechanical engineering, computer engineering, and electronics engineering. It's nearly impossible to discuss every problem's solution. We've chosen a few fundamental problems and discussed how optimization algorithms could help solve them.

Our first example problem is related to software testing and belongs to computer science. Software testing is an important activity in the software development life cycle. Software testing is used to improve software quality by identifying errors in the code. Software errors are identified with the help of test cases that are generated from the entire search space. However, identifying such test cases is difficult. It is a complex problem. In the upcoming paragraph, we talk about the test case generation problem and show how we can solve this problem before discussing the application of optimization algorithms in software testing. Let us learn more about software testing problems and point out where optimization can be done.

9.2 INTRODUCTION OF SOFTWARE TESTING

Today, many organizations are reducing the time for different operations by changing their infrastructure and installing computers in other sections. These computers support humans to complete their work in less time and reduce mistakes in computation. These days the use of computers is not limited to calculation only. Different software companies are creating application-oriented software created according to the requirement supplied by an organization. This application software is changing the working environment of the organization and is making our life very easy. Due to the very high demand for software in different areas, many companies create software for various purposes. For attracting clients, Companies are offering high-quality software at a reasonable cost. However, creating quality software is not an easy task. It is a well-defined process. This well-defined process is known as the software development life cycle.

The software development life cycle consists of many phases. The first few phases are used for developing software. Other steps are used to improve the quality of developed software and for maintaining it.

Software development starts with a feasibility study. If found feasible, its requirements are identified. After checking conditions, a top-down or bottom-up approach is followed to design the software, and coding is used to convert this design into software. After coding, software testing is used to identify errors in software and implement software as per user requirements. For performing testing of the software, two testing methods can be used. The first testing method is known as black-box testing. It is used by companies to check the bugs in the software.

In black-box testing, it is assumed that software is like a black box for the tester. A tester checks the difference between expected output and desired output by executing the test cases of the software. If there is no difference in both the output, the test case will not identify an error in the software. Such kind of test cases will not improve the quality of software. As the goal of testing is to identify the mistakes in the software. Everybody will be interested in generating those test cases that are showing the difference in both outputs. These test cases are known as practical test cases.

Another testing that is used to check software is known as white-box testing. In white-box testing, various coverage criteria are used to check the errors in software. Before performing white-box testing, it is assumed that the software has no programming errors and that the software structure is known to the tester. This structure of the program is drawn with the help of a flow graph. The programmer generates test cases that will try to cover the whole flow graph. The statement coverage, Brach coverage, and path coverage methods are used to check the coverage of the flow graph.

Let us take an example program and see how test cases can be generated. Let us write a program for adding two numbers and see how many test cases are required to check the program.

It's a C program for adding two numbers:

Example Program:

```
# include<stdio.h>
main()
{
int a, b, c;
printf("enter a and b");
scanf("%d%d", &a,&b);
c=a+b;
printf("c=", c);
}
```

9.3 TEST CASE GENERATION TECHNIQUES

Errors in software are identified by passing compelling test cases that cause the difference in expected and actual output. A test case contains a set of input variables and the expected outcome produced by software when input variables are passed in software.

Let us see how many test cases will be required to verify the correctness of the program. Our test case will be represented as follows

Input values: output value: expected output

One such test case will be

a=1 and b=1:c=2:

After running the program, if it gives 2, the program will be correct. If it is giving 1, then the program will not be correct. This was the example of one sample test case. How many test cases will be required for complete testing? Since a and b both can take any integer values, we need to check this program for all values of integers. Nearly 2^{64} test cases will be required to check the correctness of the program.

For such a simple program, the testing of all test cases will take a lot of time and will be impractical.

Ultimately, we only want to figure out those test cases which will produce the difference between exact output and desired output. If we can figure out such test cases by any means, we can reduce the errors in our software. An optimization algorithm can help us find such test cases in less time. Therefore, we are interested in learning how we can use optimization algorithms in testing.

The previous section discussed that a thoroughly tested software would have a minimum number of errors and, when delivered, will satisfy the client. However, thorough testing is not possible. It is quite impossible to check software output at all possible input values due to time and cost constraints. Even in a tiny software, the number of test cases is very large (see the example program). To improve software quality, if we are checking our software on each test case, the cost of software will increase, making a considerable loss to the software company. Also, we can't skip the testing phase as poorly tested software may result in customer dissatisfaction. If this software is designed for critical applications, it might lead to heavy financial loss or even endanger lives. Therefore, to gain confidence that software will work as per user expectations, this newly developed software must be carefully exercised by the fault revealing test cases. Searching for appropriate test data is not an easy task. It is challenging and labor-intensive work. Therefore, the test case generation problem is an optimization problem in which a set of test cases are to be searched from the extensive input domain of the variables to optimize specific coverage criteria. There is a need to generate test data automatically to save the development time and cost of the software. Other approaches used to identify helpful test cases are shown in the following subsections. However, they are not very effective in finding good test cases.

9.3.1 Random Test Data Generation Technique

This technique selects test cases randomly until useful test cases are found. But test cases generated by this technique may fail to satisfy the client's requirements as the information regarding the test requirements is not incorporated into the generation process.

9.3.2 Symbolic Test Data Generation Technique

It assigns symbolic values to the variable to create an algebraic expression for the constraints in the program. Also, it uses a constraints solver to solve the previously discussed expressions, which satisfies a test requirement. Here, in this modal, some constraints restrict the direct usage of symbolic execution. Symbolic execution cannot determine which symbolic value of the potential values will be used for array or pointer. Eventually, symbolic execution cannot find floating-point inputs as the current constraint solvers cannot solve the floating-point constraints.

9.3.3 Dynamic Test Data Generation Technique

It collects the information during the program execution to determine which test cases are best to satisfy the requirement. Then the test inputs are incrementally changed until one of them meets the requirement. As dynamic techniques depend on local search techniques such as gradient descent, these can stop or hamper the progress of testing when the local minima are encountered.

9.3.4 Metaheuristic Techniques

Metaheuristic techniques can overcome the limitations of existing approaches and generate test cases that can find hidden errors in the software. These heuristic algorithms can automatically create test cases that satisfy specific criteria, such as statement coverage, branch coverage, path coverage, and others. These techniques involve applying hill climbing, simulated annealing, PSO, and other optimization algorithms to generate test cases for a wide variety of adequacy criteria like statement coverage, branch coverage, and path coverage. Evolutionary algorithms (especially GA) are a highly cited area in this field.

This chapter proposes the application of GA and Enhanced-GA (EGA) in white-box testing. The statement coverage is taken as adequacy criteria to determine when to end the testing process. Gcov tool is a Linux-based tool that can check the statement coverage of different benchmark programs. Let us discuss some tools used to identify statement, branch, and path coverage in detail.

9.3.5 Test Case Coverage Tools

9.3.5.1 Gcov

Generating test cases that satisfy the given coverage criteria is very important for improving the quality of software. Three coverage criteria, that is, statement coverage, branch coverage, and path coverage, can be used to check the quality of test cases. Linux includes built-in tools for ensuring test case coverage. Gcov is a source code coverage tool for Linux. This service provides coverage analysis as well as statement-by-statement profiling. Gcov is a standard utility included with the GNU CC (GCC) suite that can be used to identify block and branch coverage.

More information about this tool is available on Gcov (using the GNU Compiler Collection [GCC]) website.

9.3.5.2 Trucov

Trucov is another critical tool used for checking the coverage of test cases. It is an open-source code coverage analysis tool that comes with GCC version 4.0 and later aims to be a gcov replacement. Trucov improves on gcov by providing more granular and machine-readable output, such as DOT files representing the program's control flow graph. The use of DOT Files allows for other standard tools like GraphViz to produce coverage graphs.

Trucov displays the control flow of a program and its test coverage information. It assists developers in ensuring their test cases have sufficient coverage. Trucov simplifies and formalizes how developers identify untested code in a simple-to-identify visual way similar to Gcov but better.

More information about this tool is available on Trucov—Software Testing Tools Guide.

9.4 TEST CASE GENERATION USING A GA FOR WHITE-BOX TESTING

In this section, we discuss how GAs can be used to generate test cases whose statement coverage is better than existing test cases. Since we want to create test cases with maximum coverage, this problem can be taken as an optimization problem. As we all know that it is quite impossible to complete testing of the software, we have to use an optimization algorithm that can generate required test cases in very few iterations. However, the only thing we can provide to our optimization algorithm is the coverage information as a fitness of the test case. The optimization algorithm searches for appropriate test cases by taking guidelines from previous test cases. In white-box testing, various coverage criteria are used for checking the effectiveness of the test case. Once a criterion is selected, it is used to identify the fitness of the test case. Like other problems, the algorithm starts with an initial set of test cases randomly generated from the domain of input variables. These initial sets of test cases will work as the initial population of GA. The input population will be passed through selection, crossover, and mutation operators to produce the next generation of test cases. For starting the process, the coverage of each test case is calculated with the help of a code coverage tool (for this Gcov, Lcov or Trucov) can be used.

Selection, crossover, and mutation are implemented with the help of a fitness function that will guide the search. The overall process of test case generation can be implemented with a framework in which each component will perform a separate function. Since in white-box testing, we want to cover the whole flow graph to check the correctness of the software, there will be many branches in almost all programs. So only one test case will not be able to cover the entire graph. For covering the full flow graph of the program, we have to generate several test cases. A combination of test cases that will cover a full flow graph is known as a test suite.

The framework for generating a test suite will cover these four components:

This framework has four modules:

a. Test suite generator
b. Initiator

c. Test case generator
d. Code coverage checker

9.4.1 Test Suite Generator

This module will check whether sufficient test cases have been generated. If the generated test cases can cover the entire flow graph, then the test suite will be created other new test cases will be generated again and again until the test suite is formed. The test suite generator is depicted in Figures 9.1 and 9.2. If the number of test cases is not enough to cover the whole flow graph, start the initiator.

9.4.2 Initiator

The initiator module will contain information that is required to start other modules. The following information will be available in the component:

1. Invoking by test suite generator: The initiator will be invoked by the test suite generator if the test suite is not complete. If the process is not complete, then it will start the test case generation process. If complete, it will print the test suite.

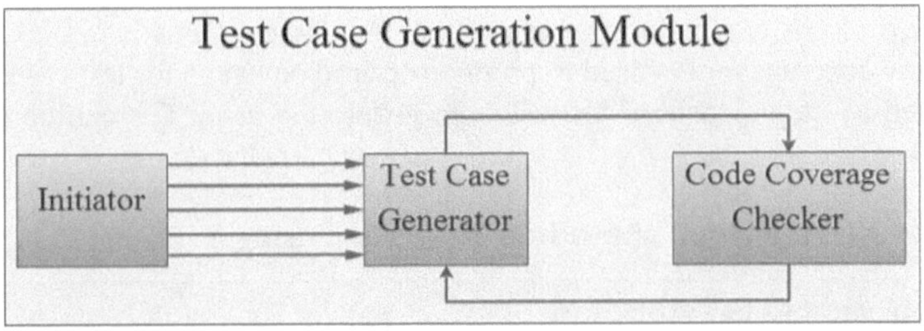

FIGURE 9.1 Test case generation module.

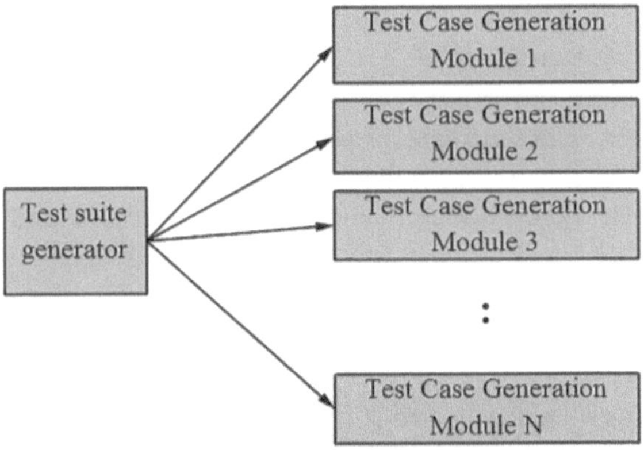

FIGURE 9.2 Test suite generator.

2. Data for starting and terminating the process: The following information will be available in this module to run test case generation. The information passed for test case generation is the maximum number of generations, population size, dimension of the solution, and type of encoding.
3. Other required information to run the algorithm: the probability of crossover, mutation, and the type of selection operator, crossover operator, and mutation operator.
4. Coverage criteria used for checking the fitness of the test case.

9.4.3 Test Case Generator Module

This module uses an optimization algorithm for generating appropriate test cases. It fetches the information from the initiator tool and generates the initial population. This initial population contains some randomly generated test cases, which are generated from the input domain. Presently, GA has been used as an optimization algorithm for test case generation. The optimization algorithm GA generates appropriate test cases using selection, crossover, and mutation operators, which have better statement coverage than existing test cases. The fitness function passed in the initiator module will assist in test case generation process.

9.4.4 Code Coverage Checker

The process of GA can be used recursively only if fitness value is available for test cases. Gcov coverage tool is used to produce desired coverage for test cases.

The step-by-step algorithm for test suite generation using GA can be explained as follows.

Algorithm 9.1: Algorithm of test suite generation using a GA

---

1. Initialize test suite `T={empty}`
2. **While** `(Coverage of Test Cases(T) != 100)`
3. Add test cases to test suite generated by GA
4. Create an initial population of size `N` by randomly selecting `N` test cases from the input domain of variables. Test cases must be represented in binary.
5. `Current generation = 1`
6. **While**`(current generation <= max Generation)`
    - 6.1. Apply Gcov and Calculate the percentage of statements covered by each test case, this will work as the fitness of each solution.
    - 6.2. Apply selection, crossover, and mutation operators and generate a new population.
    - 6.3. number of generations = number of generations + 1
7. **end while.**
8. Pick best test case `Ti`
9. Update test suite `T={T} U {Ti}`
10. **End of while (main loop)**

9.4.5 The Triangle Program

Let us take an example and explain the steps of test suite generation for white-box testing by genetic algorithm in detail. This program is a top-rated program, and it is used as a benchmark for defining many tests. This program is used to check whether the triangle is isosceles, equilateral, scalene, or not. The following example illustrates the proposed approach.

```
/*----------------------- Start of triangle program-----------------------*/
#include <stdio.h>
int main(int argc,char *argv[])
{
 float a,b,c;
 int validInput=0;
 sscanf (argv[1], "%f", &a);
 sscanf (argv[2], "%f", &b);
 sscanf (argv[3], "%f", &c);
 printf("a=%f b=%f c=%f\n",a,b,c);
if (((a+b)>c)&&((c+a)>b)&&((b+c)>a))
  validInput=1;
  else
  validInput=0;
if(validInput==1)
    {
          if((a==b)&&(b==c))
          {
          printf("The Traingle is equiletral\n");
          }
          else if((a==b)||(b==c)||(c==a))
           {
          printf("The Traingle is isosceles\n");
          }
          else
           {
           printf("The Traingle is scalene\n");
          }
    }
else if(validInput==0)
{
printf("The Values do not constitute the traingle\n");
}
return 0;
}//end of main()
/*----------------------- End of triangle program-----------------------*/
```

9.4.6 Initial Population Generation for a GA

A fixed number of test cases will work as an initial population for a GA. A test case can be formed by combining all input variables that were used to produce output. These test cases will be generated randomly from the input domain of the variables.

For example, for the test case for the triangle, the program will be a combination of variables a, b, and c, and it can be represented as abc, where a, b, and c are numerical values. Here it can be observed that the domain of these values is a closed interval of float values (0, 10), which implies that any value of a, b, and c variables within this interval will be acceptable. Since we want to apply the binary version of GA, we have to use binary encoding for variables a, b, and c. For example, if we're going to use 10 bits to represent variable a, 10 bits to represent variable b, and, finally, 10 bits for c variable, then the length of the test case will be 30 bits. Now we use a randomizer and generate random values of variables a, b, and c. Let us generate this string 1100110011 00010111011 1011111100, and then this value will be our test case. The actual value of the test case can be identified by using the Equation 9.1:

$$X = X^{min} + \frac{X^{max} - X^{min}}{2^l - 1} DV(si) \tag{9.1}$$

$DV(si)$ = Decimal Value of input string si

Binary encoding is used to encode a range of (0, 100) over 30-bit chromosomes. This 30-bit chromosome contains 10 bits for variable a, 10 bits for variable b, and 10 bits for variable c. The encoding is done according to the following formulae.

Here X^{min} = 0; X^{max} = 100; l=10; si, is the input string. For example:

The X values corresponding to the given string, 1100110011 00010111011 1011111100 are

```
Xª = (100/(1024-1))*819 = 8.005DV(1100110011)  = 819
Xᵇ = (100/(1024-1))*187 = 1.827DV(00010111011) = 187
Xᶜ = (100/(1024-1))*764 = 7.468DV(1011111100)  = 764
```

An example of the initial population follows. The size of the initial population is 20. In this example, the length of variables is taken as 7. Since the objective of the problem is to identify a set of test cases whose statement coverage is maximum, statement coverage will work as a fitness function. These test cases are passed into the Gcov coverage tool, and each test case's statement coverage is recorded. An example of the initial population and its statement coverage follows.

The sample initial population data using 7-bit numbers is tabulated in Table 9.1.

9.4.7 Selection Operator

Now selection operator is applied to produce an intermediate population. The selection process is mainly responsible for creating multiple copies of fit individuals by removing unfit solutions. Here, the roulette wheel selection method is used to generate an intermediate population. During roulette wheel selection, the statement coverage is evaluated for each test case, providing fitness values, which are then normalized. Normalization means dividing the fitness value of each individual by the sum of all fitness values so that the sum of all resulting fitness values becomes one (1). Let us

TABLE 9.1
Initial Population of Test Cases

Test Case	Fitness
{4.632162, 1.129330, 2.840551}	61.538462
{5.303110, 6.038030, 2.478816}	65.384615
{4.314577, 4.761940, 5.253757}	65.384615
{4.236579, 5.522045, 2.503537}	65.384615
{1.871886, 8.943205, 2.056095}	65.384615
{6.430898, 8.721461, 0.380711}	65.384615
{6.037396, 4.276468, 0.682581}	61.538462
{8.408456, 4.099693, 1.788924}	61.538462
{4.697391, 2.765696, 6.823011}	65.384615
{4.229282, 7.793954, 0.482156}	65.384615
{7.046056, 0.801812, 0.915676}	61.538462
{7.259335, 6.298981, 8.793895}	65.384615
{0.659427, 3.474903, 3.685602}	65.384615
{7.455550, 1.524078, 0.296683}	61.538462
{4.547011, 7.202730, 7.719748}	65.384615
{4.147516, 8.490686, 7.659512}	65.384615
{1.492558, 7.050195, 1.031176}	65.384615
{3.188893, 7.144816, 2.492452}	65.384615
{8.921877, 3.975500, 1.744095}	61.538462
{8.283277, 1.784852, 9.236363}	65.384615

TABLE 9.2
Value of Test Cases with Their Statement Coverage

Test Case	Fitness
{31.858268, 76.377953, 25.905512}	41.500000
{42.456693, 55.905512, 95.826772}	35.600000
{84.765780, 45.517885, 85.590551}	14.7000000
{39.370079, 60.629921, 66.141732}	8.2000000

demonstrate the implementation of roulette wheel selection with a small example (this example is entirely different from the preceding example and is used for a simple understanding of the roulette wheel selection operator).

Let the size of the initial population is 4, and we have generated the initial test cases. The value of these test cases with their statement coverage is shown in Table 9.2.

To make a roulette wheel, we must first normalize the fitness of each solution by dividing the statement coverage of each solution by the total statement coverage of all test cases. It will give us the proportion of each solution in the fitness wheel drawn between 0 to 1.

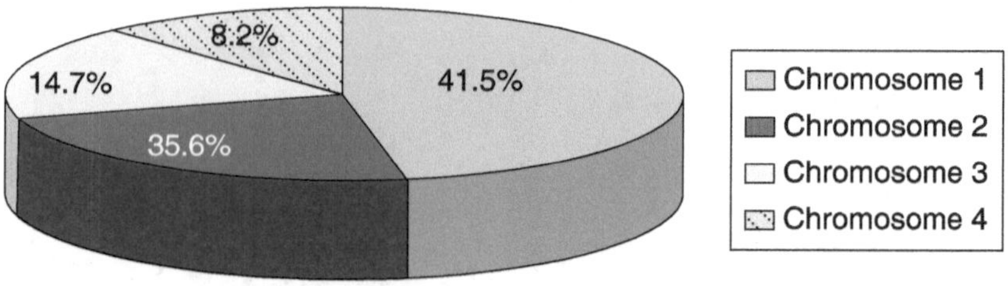

FIGURE 9.3 Roulette wheel showing fitness criteria of different chromosomes.

TABLE 9.3
Accumulated Fitness Value

Test Case	Fitness	Accumulated Fitness
{31.858268, 76.377953, 25.905512}	41.500000	0.415
{42.456693, 55.905512, 95.826772}	35.600000	0.771
{84.765780, 45.517885, 85.590551}	14.7000000	0.918
{39.370079, 60.629921, 66.141732}	8.2000000	1

From Figure 9.3, it is clear that 41.5% of the wheel will be covered by solution 1, 35.6% will be covered by solution 2, 14.7% by solution 3, and 8.2% by solution 4. Then finally accumulated normalized fitness values are computed (the accumulated fitness value of an individual is the sum of its fitness value plus the fitness values of all the previous individuals) and tabulated in Table 9.3. The earned fitness of the last individual should, of course, be 1.

Now to perform selection, we have to follow these steps:

- A random number R between 0 and 1 is chosen.
- The selected individual is the first one whose accumulated normalized value is greater than R.
    - ♣ For example, if the value of the random number is .8. Then accumulated normalized value is checked. It can be seen that this value lies in the area of solution 3; hence, solution 3 will be selected.
- These steps are repeated until a number of selected solutions become equal to the size of the initial population.

After selection, crossover and mutation are applied.

9.4.8 Crossover Operator

After selection, a crossover operator is applied to generate new test cases. To apply crossover, two test cases are selected randomly from the mating pool (the intermediate pool created after selection). Few genes of the chromosomes are exchanged between these two parents, and offspring are produced.

The simplest way to represent this is to choose some crossover points and randomly create the first child. Everything before this point should be copied from the first parent, and everything after the crossover point should be copied from the second parent. Similarly, to create a second child, everything before the crossover point should be copied from the second parent, and everything after the crossover point should be copied from the first parent. This example shows how children are generated.

Chromosome 1 11011 | 00100110110 11011 | 11000011110 child1
Chromosome 2 11011 | 11000011110 11011 | 00100110110 child2

In general, the crossover operator recombines two chromosomes, so it is also known as recombination. Crossover is an intelligent search operator that exploits the information acquired by the parent chromosomes to generate new offspring. If both the parent has the same genetic structure, then offspring are just copies of the parent irrespective of cutting point. Still, if parents have different genetic structures, then offspring are other than the parents. Thus, the crossover is the sampling process, which samples new points in search space. Generally, crossover probability is very high like 1.00, 0.95, 0.90, and so on.

9.4.9 Mutation Operator

The purpose of mutation in GAs is to preserve and introduce diversity in the solutions. The mutation prevents the algorithm from stuck in local minima by preventing the chromosomes from becoming too similar, thus slowing or even stopping evolution. The mutation operator selects one chromosome randomly from the population and then selects some genes using mutation probability and flips that bit. So mutation is a random operator that randomly alters some value. Mutation either explores some new points in the search space and leads the population to a global optimal direction or changes the value of the best chromosome and losses knowledge acquired until now. So mutation should be used rarely. Generally, per gene probability of mutation is 0.001, 0.01, 0.02, and so on.

Again a randomizer is used to select a position in the string then the bit at that position is flipped to produce offspring.

Example

Selected position (3)
Parent 11001010001101
Child 11101010001101

9.4.10 Generation of a New Population by a GA

A set of new test cases will be generated when an input population is passed to the GA. To produce a new population, each input population has to undergo selection, crossover, and mutation. Each test case will modify its input values during crossover and mutation and generate a new set of test cases. This updating will be done with the help of the fitness function (statement coverage). This process will be repeated again and again until a maximum number of generations is reached. After one iteration of selection, crossover, and mutation, the generated population was created as shown in Table 9.4.

This process has been repeated for 500 iterations and new test cases are generated to generate reasonable solutions. However, no new test case (whose coverage is more excellent than 65% is generated) as max coverage for any test case is around 65.384615. Table 9.5 shows the population was generated after 500 generations.

9.5 APPLICATION OF PSO IN TEST CASE GENERATION

Now we will discuss the application of PSO in test case generation. To explain this algorithm again, the same program is taken into consideration to create a better understanding of software testing. PSO can directly work with real-parameter encoding; hence, we can directly work with real values of test cases. There is no need to encode solutions in a binary number.

TABLE 9.4
Creation of Generated Population after One Iteration

Test Case	Fitness
{2.193995, 2.910021, 7.268741}	65.384615
{5.764022, 9.775423, 7.455101}	65.384615
{9.005824, 7.380830, 7.781671}	61.538462
{8.476796, 4.240117, 2.375222}	61.538462
{8.679434, 0.001996, 9.445435}	65.384615
{1.039493, 3.117975, 1.402183}	65.384615
{8.212260, 5.802077, 1.326413}	61.538462
{9.207221, 1.733943, 7.901189}	61.538462
{8.124159, 8.800297, 6.426503}	65.384615
{6.991039, 5.281656, 8.555809}	65.384615
{6.504221, 1.134682, 0.256120}	61.538462
{4.476240, 9.885101, 7.251243}	65.384615
{4.996655, 1.595599, 0.605815}	61.538462
{6.276550, 8.353242, 3.086356}	65.384615
{4.443199, 8.493461, 3.486482}	65.384615
{7.444193, 1.622556, 9.245911}	65.384615
{4.455531, 7.896920, 3.000768}	65.384615
{9.070433, 2.269243, 1.483108}	61.538462
{0.425779, 8.353310, 7.144715}	65.384615
{0.474032, 4.877696, 6.851752}	65.384615

TABLE 9.5
Population after 500 Iterations

Test Case	Fitness
{7.754060, 2.351343, 7.276438}	61.538462
{8.858505,2.904893, 7.148100}	61.538462
{3.275124, 6.720147, 0.901797}	65.384615
{3.179074, 1.259903, 9.042203}	65.384615
{3.238095, 1.667706, 7.004405}	65.384615
{1.011078, 9.060119, 7.563622}	65.384615
{1.462253, 7.581548, 5.094009}	65.384615
{1.284893, 3.708083, 6.441441}	65.384615
{9.294080, 4.909599, 5.895299}	61.538462
{4.731151, 7.281772, 7.484260}	65.384615
{8.411245, 1.987536, 9.282747}	65.384615
{3.917651, 3.778909, 2.096348}	61.538462
{9.063939, 9.258284, 7.942808}	65.384615
{7.200310, 2.192819, 3.075197}	61.538462
{4.411997, 4.616408, 7.094223}	65.384615
{0.037091, 4.349605, 2.730038}	65.384615
{7.988965, 7.211633, 5.940649}	61.538462

9.5.1 Particle Initialization in PSO

PSO is a population-based metaheuristic algorithm in which each particle represents a solution and moves toward better solutions in the problem search space. Here, the initial position of the particles will be initialized with the randomly generated test cases, again test cases in which the values of a, b, and c lie between 0 to 10 generated and the initial population size is 20. Similarly, the initial velocity of each particle is generated randomly from the input domain range of the variables. Initially, the test case's highest code coverage capability value will be at the `gbest` position, and its code coverage capability value will be at the `pbest` position, where `gbest` is the test case with the highest code coverage capability and `pbest` is its own best code coverage capability.

The initial positions of the particles and their fitness are shown in Table 9.6.

The initial velocities of the particles again are generated in the same way that is tabulated in Table 9.7.

9.5.1.1 Rule for Updating

In each iteration, the position of each particle/test case is updated according to the fitness function, that is, code coverage capability, so that it can reach the global optimal solution as early as possible. Here, the code coverage capability is used as the fitness function. Initial test cases with `pbest` and `gbest` values are tabulated in Table 9.8.

In the test case generation problem, there may be a situation when more than one test case achieves the same highest value of code coverage capability. In that case, any test case among them can be selected as `gbest`.

TABLE 9.6
Initial Positions of the Particles and Their Fitness

Test Case	Fitness
{9.640122, 4.269078, 8.080854}	61.538462
{0.500125, 6.260038, 2.736861}	65.384615
{8.198715, 8.160112, 6.408052}	61.538462
{9.857150, 7.792352, 4.511664}	61.538462
{6.302793, 8.652542, 8.555883}	65.384615
{7.292461, 8.820163, 9.169045}	65.384615
{1.573152, 3.792930, 9.142013}	65.384615
{1.378707, 1.550871, 8.075872}	65.384615
{2.589187, 4.794397, 3.236929}	65.384615
{6.150882, 2.145515, 5.358735}	61.538462
{0.192346, 2.310977, 6.487457}	65.384615
{9.131547, 7.898085, 3.526920}	61.538462
{1.929179, 7.761554, 1.380231}	65.384615
{8.546165, 8.699397, 8.056184}	65.384615
{9.891034, 6.078202, 4.177108}	61.538462
{7.543240, 4.761038, 6.113518}	61.538462
{5.883856, 8.277732, 3.460229}	65.384615
{9.181293, 9.548669, 6.159396}	65.384615
{2.811057, 0.995814, 4.414712}	65.384615
{2.769413, 4.848817, 3.456165}	65.384615

TABLE 9.7
Initial Velocities of the Particles

{8.818898, 9.055118, 6.456693}
{3.228346, 1.732283, 6.771654}
{4.960630, 4.566929, 5.826772}
{0.472441, 4.724409, 9.527559}
{1.102362, 6.062992, 5.196850}
{2.047244, 7.322835, 2.677165}
{4.409449, 3.700787, 9.212598}
{5.669291, 7.559055, 2.047244}
{3.149606, 7.086614, 4.173228}
{1.417323, 5.275591, 0.629921}
{4.094488, 4.488189, 3.149606}
{1.181102, 0.866142, 7.244094}
{7.637795, 0.314961, 8.503937}
{4.960630, 4.881890, 7.322835}
{2.598425, 0.393701, 8.740157}
{6.535433, 2.913386, 8.661417}

TABLE 9.7 Continued

{5.590551, 7.007874, 8.346457}
{3.622047, 7.007874, 1.653543}
{0.000000, 7.086614, 0.393701}
{2.204724, 0.787402, 8.582677}

TABLE 9.8
Initial Test Cases with pbest and gbest

Position of Particles	Fitness	pbest	gbest
{9.640122, 4.269078, 8.080854}	61.538462	{9.640122, 4.269078, 8.080854}	
{0.500125, 6.260038, 2.736861}	65.384615	{0.500125, 6.260038, 2.736861}	
{8.198715, 8.160112, 6.408052}	61.538462	{8.198715, 8.160112, 6.408052}	
{9.857150, 7.792352, 4.511664}	61.538462	{9.857150, 7.792352, 4.511664}	
{6.302793, 8.652542, 8.555883}	65.384615	{6.302793, 8.652542, 8.555883}	
{7.292461, 8.820163, 9.169045}	65.384615	{7.292461, 8.820163, 9.169045}	
{1.573152, 3.792930, 9.142013}	65.384615	{1.573152, 3.792930, 9.142013}	
{1.378707, 1.550871, 8.075872}	65.384615	{1.378707, 1.550871, 8.075872}	
{2.589187, 4.794397, 3.236929}	65.384615	{2.589187, 4.794397, 3.236929}	
{6.150882, 2.145515, 5.358735}	61.538462	{6.150882, 2.145515, 5.358735}	**{9.181293, 9.548669, 6.159396}**
{0.192346, 2.310977, 6.487457}	65.384615	{0.192346, 2.310977, 6.487457}	
{9.131547, 7.898085, 3.526920}	61.538462	{9.131547, 7.898085, 3.526920}	
{1.929179, 7.761554, 1.380231}	65.384615	{1.929179, 7.761554, 1.380231}	
{8.546165, 8.699397, 8.056184}	65.384615	{8.546165, 8.699397, 8.056184}	
{9.891034, 6.078202, 4.177108}	61.538462	{9.891034, 6.078202, 4.177108}	
{7.543240, 4.761038, 6.113518}	61.538462	{7.543240, 4.761038, 6.113518}	
{5.883856, 8.277732, 3.460229}	65.384615	{5.883856, 8.277732, 3.460229}	
{9.181293, 9.548669, 6.159396}	**65.384615**	**{9.181293, 9.548669, 6.159396}**	
{2.811057, 0.995814, 4.414712}	65.384615	{2.811057, 0.995814, 4.414712}	
{2.769413, 4.848817, 3.456165}	65.384615	{2.769413, 4.848817, 3.456165}	

Note: The values in bold is the global best solution.

Before updating each test case's velocity and position, we first have to generate random values of r_1 and r_2, let's say `0.4` and `0.2`, respectively, as these are the random numbers in the range `[0, 1]`. Values of inertia weight ω, c_1 and c_2 are taken as $\omega = 0.72$, $c_1 =$ `2.5` and $c_2 =$ `2.5`.

Now, the velocity and position of each particle are updated by using Equations 9.2 and 9.3:

$$V_i = \omega * V_i + c_1 * r_1 * (P_i - X_i) + c_2 * r_2 * (G - X_i) \tag{9.2}$$

$$X_i = X_i + V_i \ldots\ldots\ldots\ldots \quad (9.3)$$

For test cases T1, T2, T3, T4, and T6, in the first iteration, the velocity is updated according to the Equation 9.2:

```
v11 = 0.72*(8.818898,9.055118,6.456693)+ 2.5*0.4((9.640122,
      4.269078,8.080854)-(9.640122,4.269078,8.080854))+
      2.5*0.2((9.181293,9.548669,6.159396)-
      (9.640122,4.269078,8.080854))
=6.12019206, 9.15948046, 3.68808996
```

Now the position of test case T1 is updated according to Equation 9.3:

```
=(9.640122, 4.269078 , 8.080854)+ (v1)
=(9.640122, 4.269078 , 8.080854)+(6.12019206,9.15948046,3.6880
 8996)
=(15.76031406, 13.42855846, 11.76894396)
```

For a search space bounded by `[0, 10]`, position clamping is required to limit the position to the range `[0, 10]`. So, the new position of the particle will be `(10, 10, 10)`.

Similarly, the velocity and position of T2, T3, T4, and T6 are calculated. The updated positions of the particles are shown in Table 9.9.

TABLE 9.9
Updated Positions

Position of Particles	Fitness
{10, 10, 10}	65.384615
{3.999761, 7.561268, 8.121564}	65.384615
{2.157511, 9.544250, 9.586927}	65.384615
{9.977901, 8.740032, 9.272726}	61.538462
{8.883111, 6.034111, 7.293412}	61.538462
{4.862334, 8.685352, 2.719685}	65.384615
{8.890265, 5.259121, 0.356240}	61.538462
{3.156292, 9.505225, 2.934368}	65.384615
{5.357705, 1.524658, 5.568763}	65.384615
{4.529696, 9.371738, 5.903094}	65.384615
{2.244097, 5.839787, 1.804119}	65.384615
{0.511101, 4.569671, 6.513992}	65.384615
{8.455724, 2.199260, 4.896650}	61.538462
{3.031728, 9.509347, 6.664356}	65.384615
{5.124193, 8.874113, 7.315469}	65.384615
{7.844703, 6.390291, 4.807649}	61.538462
{7.716951, 3.685710, 8.466443}	65.384615
{9.831614, 2.488844, 8.449412}	61.538462

In Table 9.10, X_i is the initial position of the test case, while X_1 is the updated position after the first iteration. Fitness is calculated as Fitness(X_1) for each updated position by using Equation 9.3. The highest fitness value of a test case between initial and first iteration is selected as `pbest` and test case that has the highest value of fitness among `pbest` is taken as `gbest`.

Similarly, in the second iteration, the value of ω is updated as

$$\omega = 0.4.$$

Now the velocities and positions of each test case are updated.

Here `pbest` contains the generated test cases with higher code coverage capability than the initial test cases. This example illustrates the process of test case generation up to one iteration. The exact process is repeated for 500 iterations.

Algorithms and Results:

In this section, the PSO algorithm has been applied to generate test cases for generating a test case that has higher code coverage capability as compared to the initial modification traversing.

In the proposed approach, *Gcov* Tool is used to evaluate the test cases' statement coverage. A C language program has been developed to generate new test cases by taking the statement coverage information as a fitness value of the test cases from the *Gcov* file.

The stepwise algorithm is as follows:

Algorithm 9.2: Algorithm of Test Suite Generation Using a PSO Algorithm

---

1. These test cases are initialized as an input swarm to the PSO algorithm.

2. Initial velocity is randomly generated in the range of the input domain.

3. The values of ω, c_1, c_2, r_1, and r_2 are declared.

4. The `pbest` and `gbest` for the initial iteration are selected according to the calculated code coverage capability of the input swarm.

5. In each iteration, new test cases are generated by the update rule of the proposed algorithm.

6. The fitness value of each generated test case is calculated according to Equation 9.3.

7. Steps 5 and 6 are repeated until a termination criterion is met.

8. The average code coverage capability score for the generated test suite is calculated by using equation 9.4.

9. The termination criteria are based on "until the maximum number of iterations is achieved or until acceptable average code coverage capability score is achieved."

10. To evaluate the proposed approach, individual and average code coverage capabilities of the generated test cases are compared with that of initial test cases.

TABLE 9.10
Updated Positions

X_i	Fitness(X_i)	X_1	Fitness(X_1)	Pbest	gbest
{9.640122, 4.269078, 8.080854}	61.538462	{10, 10, 10}	65.384615	{10, 10, 10}	
{0.500125, 6.260038, 2.736861}	65.384615	{3.999761, 7.561268, 8.121564}	65.384615	{3.999761, 7.561268, 8.121564}	
{8.198715, 8.160112, 6.408052}	61.538462	{2.157511, 9.544250, 9.586927}	65.384615	{2.157511, 9.544250, 9.586927}	
{9.857150, 7.792352, 4.511664}	61.538462	{9.977901, 8.740032, 9.272726}	61.538462	{9.977901, 8.740032, 9.272726}	
{6.302793, 8.652542, 8.555883}	65.384615	{8.883111, 6.034111, 7.293412}	61.538462	{6.302793, 8.652542, 8.555883}	
{7.292461, 8.820163, 9.169045}	65.384615	{4.862334, 8.685352, 2.719685}	65.384615	{4.862334, 8.685352, 2.719685}	
{1.573152, 3.792930, 9.142013}	65.384615	{8.890265, 5.259121, 0.356240}	61.538462	{1.573152, 3.792930, 9.142013}	
{1.378707, 1.550871, 8.075872}	65.384615	{3.156292, 9.505225, 2.934368}	65.384615	{3.156292, 9.505225, 2.934368}	
{2.589187, 4.794397, 3.236929}	65.384615	{5.357705, 1.524658, 5.568763}	65.384615	{5.357705, 1.524658, 5.568763}	
{6.150882, 2.145515, 5.358735}	61.538462	{4.529696, 9.371738, 5.903094}	65.384615	{4.529696, 9.371738, 5.903094}	{0.511101, 4.569671, 6.513992}
{0.192346, 2.310977, 6.487457}	65.384615	{2.244097, 5.839787, 1.804119}	65.384615	{2.244097, 5.839787, 1.804119}	
{9.131547, 7.898085, 3.526920}	61.538462	**{0.511101, 4.569671, 6.513992}**	**65.384615**	{0.511101, 4.569671, 6.513992}	
{1.929179, 7.761554, 1.380231}	65.384615	{8.455724, 2.199260, 4.896650}	61.538462	{1.929179, 7.761554, 1.380231}	
{8.546165, 8.699397, 8.056184}	65.384615	{3.031728, 9.509347, 6.664356}	65.384615	{3.031728, 9.509347, 6.664356}	
{9.891034, 6.078202, 4.177108}	61.538462	{5.124193, 8.874113, 7.315469}	65.384615	{5.124193, 8.874113, 7.315469}	
{7.543240, 4.761038, 6.113518}	61.538462	{7.844703, 6.390291, 4.807649}	61.538462	{7.844703, 6.390291, 4.807649}	
{5.883856, 8.277732, 3.460229}	65.384615	{7.716951, 3.685710, 8.466443}	65.384615	{7.716951, 3.685710, 8.466443}	
{9.181293, 9.548669, 6.159396}	65.384615	{9.831614, 2.488844, 8.449412}	61.538462	{9.181293, 9.548669, 6.159396}	
{2.811057, 0.995814, 4.414712}	65.384615	{10, 10, 10}	65.384615	{10,10, 10}	
{2.769413, 4.848817, 3.456165}	65.384615	{3.999761, 7.561268, 8.121564}	65.384615	{3.999761, 7.561268, 8.121564}	

Note: The values in bold depicts pbest.

9.6 APPLICATION OF DE IN TEST CASE GENERATION

Like PSO, DE can also be used in test case generation. Moreover, it can directly work with real-parameter encoding; hence, it is unnecessary to encode solutions in a binary number.

9.6.1 Initialization Population Generation in DE

Similar to GAs and PSO, the initialization of 20 solutions of three-dimensional real-valued parameters vectors with the random values between the range 0 and 10 will be done, where $x_{j,min}$ will be 0 and $x_{j,max}$ will be 10. $\texttt{rand}_{i,j}\texttt{[0,1]}$ will be a random value that will be in between 0 to 1, and D will be 3.

$$X_{j,i,0} = X_{j,min} + rand_{i,j}\left[0,1\right] * \left(X_{j,max} - X_{j,min}\right), \tag{9.4}$$

where $1 \le i \le 20$
$1 \le j \le D$
$0 < rand_{i,j}[0,1] < 1.$

Twenty randomly generated test cases in which the values of a, b, and c lie between 1 to 10 will work as the initial population. These solutions are evaluated, and the statement coverage of each solution will work as fitness.

Initial population with fitness is listed in Table 9.11.

TABLE 9.11
Initial Population with Fitness

Test Case	Fitness
{2.204724, 0.787402, 8.582677}	65.384615
{0.000000, 4.566929, 5.511811}	65.384615
{4.803150, 1.889764, 9.212598}	65.384615
{3.858268, 1.181102, 6.535433}	65.384615
{5.039370, 7.401575, 4.330709}	65.384615
{5.826772, 4.094488, 9.370079}	65.384615
{3.228346, 1.889764, 5.354331}	65.384615
{6.141732, 9.527559, 5.748031}	65.384615
{2.913386, 3.543307, 2.440945}	65.384615
{3.543307, 7.165354, 7.716535}	65.384615
{2.992126, 9.133858, 5.590551}	65.384615
{3.622047, 7.165354, 3.149606}	65.384615
{0.472441, 6.062992, 0.629921}	65.384615
{4.094488, 7.244094, 3.464567}	65.384615
{7.480315, 2.519685, 4.094488}	61.538462
{9.370079, 2.913386, 3.070866}	61.538462
{8.425197, 1.653543, 4.960630}	61.538462
{7.795276, 0.708661, 4.881890}	61.538462
{9.606299, 7.401575, 5.354331}	61.538462
{8.188976, 2.755906, 4.645669}	61.538462

In each iteration, each solution/test case has to pass through mutation and crossover operator.

9.6.2 Mutation

A donor vector is created during mutation by choosing three initial vectors, also known as target vectors. First, a vector known as a target vector is chosen from the current generation and added to the scaled difference of two other vectors, where r_1, r_2, and r_3 are the positions of vectors and i is the position of the mutated vector. For example, if we want to create a mutation vector for first solution, we will randomly select three vectors from the pool of `20` solutions. If the positions of those vectors are 3, 4, and 5, then vector `(4.803150, 1.889764, 9.212598)` will work as $\vec{X}$ `31,0`, `(3.858268, 1.181102, 6.535433)` will be $\vec{X}$ `41,0` and `(5.039370, 7.401575, 4.330709)` will be $\vec{X}$ `51,0`. With the help of Equation 9.5, we will identify the value of the donor vector for the first solution. The donor vectors are tabulated in Table 9.12.

$$\vec{V}_{i,G} = \vec{X}_{r1^i,G} + F * \left(\vec{X}_{r2^i,G} - \vec{X}_{r3^i,G} \right), \tag{9.5}$$

TABLE 9.12
Donor Vectors

Donor Vectors
{7.257116, 9.201374, 2.420238}
{2.601631, 3.333664, 6.644838}
{1.031448, 7.275416, 6.138943}
{2.900682, 2.608916, 0.194975}
{0.219347, 9.498149, 5.299300}
{9.123309, 2.990903, 1.998043}
{8.294872, 5.942238, 4.985943}
{1.864070, 9.448130, 2.680643}
{0.501398, 2.596907, 4.026530}
{4.381190, 1.432939, 8.319282}
{0.646193, 2.701666, 2.370552}
{8.159851, 7.168738, 5.433551}
{5.384682, 3.605857, 2.993724}
{8.086622, 1.687247, 8.798648}
{4.755005, 4.746485, 8.588671}
{7.722852, 6.306125, 3.787017}
{2.128726, 3.778740, 8.254570}
{7.174023, 1.484035, 5.466517}
{5.844690, 9.087323, 0.656873}
{7.827836, 1.390544, 5.959688}

where $\vec{V}_{i,G}$ is the donor vector.

$\vec{X}_{r1^i,G}$, $\vec{X}_{r2^i,G}$ and $\vec{X}_{r3^i,G}$ are three chosen vectors from current generation G.

F is scalar number [0.4 - 1.0].

$r_1 \neq r_2 \neq r_3$ are mutually exclusive.

We will repeat this process for all 20 solutions, and finally, this population is generated.

9.6.3 Crossover

In order to enhance diversity within the population crossing of the donor vector $\vec{V}_{i,G}$, with target vector $\vec{X}_{i,G}$, is done where their components are exchanged to form trial vector $\vec{U}_{i,G}$. The trial vector is computed by Equation 9.6:

$$\vec{U}_{j,i,G} = \vec{V}_{j,i,G} \text{ if}\left(rand_{i,j}\left[0,1\right] \leq Cr \text{ or } j = j_{rand}\right)$$

$$\vec{X}_{j,i,G} \text{ otherwise,} \tag{9.6}$$

where $0 < rand_{i,j}$ [0,1] <1

Cr is crossover rate, and

j_{rand} ε [1,2, . . . , D].

For example, to create a trial vector for the first solution, we will take the first donor vector from the donor vectors, that is, (7.257116, 9.201374, 2.420238), and combine it with the corresponding target vector, that is, (2.204724, 0.787402, 8.582677). Since in testing, the crossover rate is taken as low because we do not need much diversity. We will choose the value of *Cr* accordingly. Let the value of *Cr* is 0.4. Now for each dimension, we will generate the random values in between 0 to 1. If the generated values are (0.1, 0.3, 0.2), then the generated trial vector will have only the components of the donor vector. Let us check that the first random number it is 0.1 when we compare this value with *Cr*. This will choose the element from the donor vector. A similar process, we repeat for other values of the vector. Since all random values are less than *Cr*, the trial vector remains the same as the donor vector. When this process is applied for all the vectors, these trial vectors are generated. The output in the form of the test case and the fitness values for the trial vector are depicted in Table 9.13.

9.6.4 Selection Operator

The selection operator applies the survival of the fittest principle to determine the winner of the race for the next generation. The fitness values of the target vector and the trial vector are compared, and the best one is selected. Because the goal of the test case generation problem is to generate a vector with a higher fitness value, the test case with the most increased fitness is always chosen. Equations 9.7 and 9.8 will select the candidate for this problem, and Table 9.14 shows the corresponding results.

TABLE 9.13
Test Case and Fitness Values for Trial Vector

Test Case	Fitness
{7.257116, 9.201374, 2.20238}	65.384615
{2.601631, 3.333664, 6.644838}	65.384615
{1.031448, 7.275416, 6.138943}	65.384615
{3.858268, 1.181102, 6.535433}	65.384615
{0.219347, 9.498149, 5.299300}	65.384615
{5.039370, 7.401575, 4.330709}	65.384615
{5.826772, 4.094488, 9.370079}	65.384615
{1.864070, 9.448130, 2.680643}	65.384615
{0.501398, 2.596907, 4.026530}	65.384615
{4.381190, 1.432939, 8.319282}	65.384615
{0.646193, 2.701666, 2.370552}	65.384615
{3.622047, 7.165354, 3.149606}	65.384615
{0.472441, 6.062992, 0.629921}	65.384615
{8.086622, 1.687247, 8.798648}	65.384615
{4.755005, 4.746485, 8.588671}	65.384615
{7.722852, 6.306125, 3.787017}	61.538462
{2.128726, 3.778740, 8.254570}	65.384615
{7.174023, 0.708661, 5.466517}	61.538462
{5.844690, 9.087323, 0.656873}	65.384615
{7.827836, 1.390544, 4.645669}	61.538462

$$\vec{X}_{i,G+1} = \vec{U}_{i,G}\ , if\ \ f\left(\vec{U}_{i,G}\right) \geq f\left(\vec{X}_{i,G}\right) \tag{9.7}$$

and

$$\vec{X}_{i,G}\ \ if\ f\left(\vec{U}_{i,G}\right) < f\left(\vec{X}_{i,G}\right), \tag{9.8}$$

where *f*(*X*) is the fitness function to be maximized.

Again, the same process is repeated for 500 iterations, and the final population is generated.

9.7 CONCLUSION

This chapter proposes how GA and PSO generate the testing test suites for white-box testing to achieve the maximum statement coverage. As discussed in the previous chapter, these two algorithms efficiently produce an optimal result for any problem. We have exploited these algorithms to create the test data.

TABLE 9.14
Test Case and Fitness Values for the Trial Vectors after the Selection Operator

Test Case	Fitness
{7.257116, 9.201374, 2.20238}	65.384615
{2.601631, 3.333664, 6.644838}	65.384615
{1.031448, 7.275416, 6.138943}	65.384615
{2.900682, 2.608916, 0.194975}	61.538462
{0.219347, 9.498149, 5.299300}	65.384615
{9.123309, 7.401575, 4.330709}	61.538462
{8.294872, 5.942238, 4.985943}	61.538462
{1.864070, 9.448130, 2.680643}	65.384615
{0.501398, 2.596907, 4.026530}	65.384615
{4.381190, 1.432939, 8.319282}	65.384615
{0.646193, 2.701666, 2.370552}	65.384615
{8.159851, 7.168738, 5.433551}	61.538462
{5.384682, 3.605857, 2.993724}	61.538462
{8.086622, 1.687247, 8.798648}	65.384615
{4.755005, 4.746485, 8.588671}	65.384615
{7.722852, 6.306125, 3.787017}	61.538462
{2.128726, 3.778740, 8.254570}	65.384615
{7.174023, 0.708661, 5.466517}	61.538462
{5.844690, 9.087323, 0.656873}	65.384615
{7.827836, 1.390544, 4.645669}	61.538462

10 Application of Genetic Algorithms, Partial Swarm Optimization, and Differential Evolution in Regression Testing

10.1 INTRODUCTION

In the previous chapter, we have discussed why software testing is necessary. What will happen if the software developer does not perform testing of the software? In the feasibility study, the software developer estimates the time and cost of developing the software. Before creating the software, the software developer creates a document where both developer and client sign a document containing the cost and time for developing the software. After getting approval from the client, the software developer starts the development of the software product. During software development, they have to watch the time and money invested and plan the action accordingly. Before testing, the software tester will check how much money and time they have in hand for the project? After checking that information, they start testing. If a minimal budget is available for testing, the software tester performs minimum testing by designing and developing test cases per requirement specifications. By running these test cases, they may be sure that they have already tested all those test cases, which the end users will frequently feed. However, if budget and time permit, they can generate some more test cases with the help of an optimization algorithm. After completing the testing process, the company will deliver the software product to the client for further use. More testing will remove more errors from the software and make it more reliable.

10.2 REGRESSION TESTING

The client uses software for their work. As time passes, due to improper use of the software, some errors may occur in the software. Sometimes clients want to add more functionality to the existing software. A software developer can fix these issues either by repairing or changing software functionality during the maintenance phase.

DOI: 10.1201/9781003313649-10

After setting the software problems, the software developer must retest the software to verify that newly developed software is not having errors. This retesting of software is known as regression testing. When we do regression testing, we ensure that a change to an existing section does not impact the other units and the newly created version is correct as per new requirements.

As we know that the purpose of regression testing involves testing a modified version of the original software; hence, developers do not want to consume a lot of time on it. They wish to deliver this software to the client as soon as possible. However, developers still want to test frequently used code at minimum cost to ensure that the new version is correct. Therefore, in place of regenerating new test cases for the latest version of the software, the tester minimizes the time and cost of regression testing by developing a new test suite for the latest version of the software by updating an already available test suite created for preexisting software. The tester only generates test cases that only cover the changed portion of the code and add these to the pre-existing test suite. However, this will increase the size of the test suite. To reduce the size of the test suite, the software tester reruns the original test cases on a new version of the software, keeping only the most useful test cases. Cases that are redundant or obsolete are removed. When you execute the test cases of a new test suite on newly updated software, some test cases will be useful, while others will not. Because this is modified software, not all existing test cases will be rendered ineffective. Some test cases will continue to run in the new version. Obsolete test cases are existing test cases that are no longer applicable in the most recent version. However, before running this new test suite on a new version of the program, we can identify test cases that are redundant or obsolete. We employ several techniques to reduce the size of the test suite. Several steps are involved in the process of analyzing the existing test suite and refining it to make it suitable for rerun. The analysis of an updated test suite entails three critical steps: minimization, prioritization, and test case selection.

Test Suite Minimization:

The minimization process of the test case involves reducing the size of the test suite by removing obsolete and redundant ones.

Once the new software version is created, the Software tester will generate new test cases to test the changed code. After adding new test cases to the existing ones, we receive aggregated test suits suitable for updated software. However, doing so will increase the size of test cases meant for the newer version. This new test suite will have three kinds of test cases, that is, (1) obsolete test cases, (2) functional test cases, and (3) new test cases.

10.2.1 Test Suite Minimization Problem

Test suite minimization refers to reducing the size of the test suite by removing redundant test cases:

- A test suite $T = \{t_1; t_2; \ldots ;t_i\}$ and a set of test requirements $R= \{r_1; r_2; \ldots ; r_j\}$.
- To find the minor T′ such that $T' < T$, for all r belongs to R (T′ satisfies r).

We know there will be some cost involved when we minimize the test cases. We also know that there will be a reduction in the cost involved in the testing phase because the size of the test suite has been decreased. Hence, minimizing the test cases will be profitable, or you can say effective only if

- the cost involved in reduction of test cases < gain achieved by testing with reduced test suit.

A drawback of test suite minimization is that when we decrease the number of test cases, our test suite might become ineffective to detect the fault in the software.

10.2.2 Test Case Selection

This is another method to decrease the size of the test suit. As the name suggests, it works on the principle of selective decrement where a software tester only looks into the modifications of the code. Based on the modifications, it traverses only that part of the test cases, which are relevant to the modification. So, it can be related to the white-box testing in which the tester is exposed to the internal working of the code. Several methods are present for test case selection. Among them, most of the techniques are based on

- adaptive random testing,
- control flow graph,
- genetic algorithms,
- greedy algorithms, and
- dependence graphs.

10.2.2.1 Test Case Selection Problem

- The program P, modified version of P, P′ and a test suite, T.
- Find a subset of T, T′ worth which to test P′.

10.2.3 Test Case Prioritization

In the pool of test cases, there may be some cases that are related to the boundary conditions of the domain to which our code/software responds. Apart from that, we may also have some critical cases whose output needs to be checked foremost. Those test cases should be given a higher priority than the rest. Test case prioritization deals with this situation where we give priority to test cases and convert the sequence of test cases as an ordered one. We also focus on the recent code changes in the software while prioritizing. The following two capabilities need to be maintained while we prioritize our test cases:

1. Fault detection capability
2. Code coverage capability

10.2.3.1 Test Case Prioritization Problem

- A test suite T. The set of permutations of T, PT, and a function to real numbers, f: PT->R.

10.2.4 Application of a Genetic Algorithm in Test Suite Minimization

The problem of finding the minimal set of test cases that satisfies the ideal set requirements can be classified as an NP-complete problem. As we know, a suitable heuristic algorithm can solve an NP-complete problem. Here we discuss the application of a genetic algorithm (GA) in a test suite minimization problem. A GA can be used to test software over a subset of the test suite that will fulfill our requirement. Doing this will reduce our effort as well as cost. Our first concern will be to choose that subset that is possibly the best among all the test cases. For such a selection GA provides two operators crossover and mutation. Choosing a GA will fulfill all three requirements of test cases to prioritize, select, and minimize.

For example, suppose we have a problem in which given three sides, we have to find the feasibility of a triangle formation. Consider the size of the test suite to be 1000, then the total number of feasible test cases would be 2^{1000}. It is ideally not possible to test all 2^{1000} test cases.

In spite of that, if we want to test without degrading the fault detecting capability of the software then the use of a GA justifies, as it will select the best test cases using an initial population and its fitness before applying crossover operator to generate a set of new test cases.

10.2.4.1 Minimization of Pre-Prioritized Test Suite Using a GA

For minimizing the pre-prioritized test suite GA selects a subset of test suite in such a way so that the branch coverage remains the same as the original test suite. For the minimization of pre-prioritized test suite, it should go through the three modules (starting module, minimization module, and test case coverage tool) of the framework (see Figure 10.1).

10.2.4.1.1 Starting Module

It contains all the information needed to start successor modules. Contents of this module can be listed as follows:

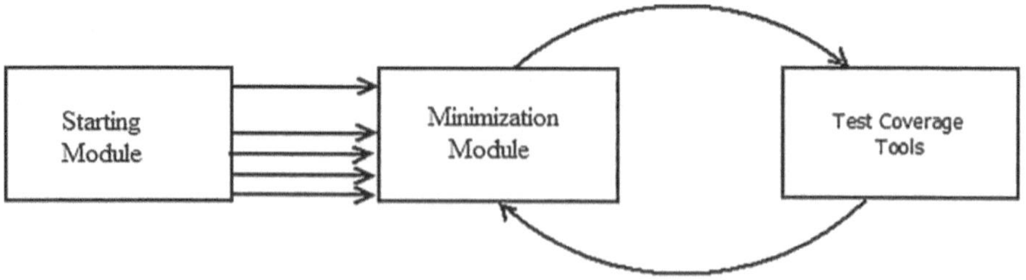

FIGURE 10.1 Test case minimization using GA.

1. Prerequisites to start GA:
   1.1. Total number of solutions in the population
   1.2. Max number of generations
   1.3. Type of used encoding
2. Parameters required to produce the next-generation population:
   2.1. Total number of test data for every individual
   2.2. Data type of input variable
   2.3. Values of parameters related to the population; in this case of solving by a GA, it may store the probability of doing crossover and mutation.
3. Fitness function used to evaluate the fitness of individual
4. Information about test tool

10.2.4.1.2 Minimization Module

In this module's initial population, the first generation of the population is generated by using the required operators. It can be said that this module is responsible for implementing the functionality of a GA.

10.2.4.1.3 Test Case Coverage Module

This module is responsible for checking the modules of the code and its corresponding test cases for the overall coverage of the software.

A GA helps in searching appropriate test subsets, having better branch coverage as compared to existing subsets of test cases. In this module, the chosen fitness function represents the branch coverage of a particular subset of test cases.

We have a plethora of coverage tools available. Among them we will be more interested to use that one that has proven quality and open source.

Gcov is one such coverage tool (you can find it here: *http://gcc.gnu.org/onlinedocs/gcc/Gcov.html*), which is used to calculate the amount of branch covered in percentages for every test case.

The following pseudocode explains the process through which the minimization of the test suite is achieved.

1. Randomly select T test subsets from the preexisting test suite to get an initial population of size T. Constraints: The test subsets should be in binary.
2. Conditional loop (number of generations < maximum number of generations)
   2.1 Apply Gcov and find out the coverage in terms of percentage by every test subset, which will be treated as its fitness.
   2.2 Generate a new population by using selection, crossover, and mutation.
3. End loop
4. Print: Final sub set.

10.2.5 Example Program

Let us work with the same example which we have used in previous chapter of test case generation. Assume we have already developed a test suite. The values are of test cases are ({1 4 8}, {7 5 0}, {0 5 10}, {8 6 6}, {2 2 10}, {8 8 2}, {8 8 9}, {10 8}, {2 2 3}, {2 3 4}, {9 3 4}, {4 4 1}) and branch coverage of each test case is known. We have modified the program and now we want to check which subset of this test suite will cover maximum branches.

10.2.5.1 Initial Population Generation for a GA

The initial population of GA will consist of a set of randomly generated test cases. The following discussion will show you how the initial population is being generated. Suppose we have a test suit, TS, comprising n number of test cases. Hence, it can be represented as TS = {1τ1, 1τ2, . . . , 1τn}; to generate initial population, we have to decide some order in which we will execute these test cases. Let 1τ1, 1τ2, . . . , 1τn. be our ordered set so that the testing will first choose 1τ1 test case, then 1τ2, and so on. Once we have fixed this ordering, we can generate the initial population as follows. We will generate a subset of this ordered set randomly and represent this subset in binary as follows. We assign 0 and 1 according to the presence of the test case in the selected subset of test case; that is, if a particular test case is present in the selected subset of test case, we assign 1; otherwise, 0.

In our example, if a subset is generated {1τ3, 1τ4, 1τ5, 1τ9} from an ordered set of 12 test cases, then it will be represented in binary as 001110001000. Hence, it can be inferred that our test set would be {1τ1', 1τ2', 1τ3, 1τ4, 1τ5, 1τ6', 1τ7', 1τ8', 1τ9, 1τ10', 1τ11', 1τ12'}. Here 1τ represents a test case that is present, and 1τ' represents the absence of that test case in this test suite. So 1τ1' represents that 1τ1 is not there. Here, a value of 1 shows the presence and 0 the absence of that particular test case. The results of test case generation phase for the initial population by elitist GA for a triangle problem is shown in Table 10.1.

10.2.5.2 Selection for a GA

The process of selection is fitness proportionate where we select potentially useful solutions, and for that, we choose a roulette wheel approach. These selected test subsets are used for a future breeding process, like crossover and mutation. After the selection process, the selected test subsets are given in Table 10.2.

10.2.6 Crossover for a GA

In the proposed algorithm, we use uniform crossover. Uniform crossover uses a fixed mixing ratio between two parents. Uniform crossover enables the parent chromosomes to contribute at the gene level rather than at the segment level. The results of the crossover operator are shown in Table 10.3.

10.2.7 Mutation for a GA

After the crossover operator, a mutation operator has been implemented in the process of the GA. The crossover operator provides the exploration as well as exploitation performance to the GA. The results of the crossover operator are shown in Table 10.4.

10.2.8 Generation of a New Population by a GA

A new and fresh generation of candidate solutions is formed after selection, crossover, and mutation operation. The results are shown in Table 10.5.

TABLE 10.1
Test Case Generation Phase for the Initial Population by an Elitist GA for a Triangle Problem

Test Subset	Binary Representation of Test Suite	Fitness (branch coverage of the subset)
{τ3,τ5,τ7,τ8,τ11}	001010110010	84.000000
{τ1,τ2,τ3,τ5,τ6,τ9,τ10,τ11}	111011001110	84.000000
{τ1,τ4,τ6,τ7,τ9}	100101101000	92.000000
{τ2,τ5,τ12}	010010000001	76.000000
{τ1,τ4,τ5,τ6,τ7,τ8,τ9,τ10,τ11}	100111111110	92.000000
{τ1,τ2,τ6,τ9,τ10,τ12}	110001001101	84.000000
{τ1,τ6,τ8,τ9,τ11}	100001011010	84.000000
{τ4,τ5,τ8,τ9,τ10,τ11,τ12}	000110011111	92.000000
{τ2,τ3,τ4,τ5,τ6,τ8,τ9,τ10,τ11}	011111011110	92.000000
{τ1,τ3,τ4,τ5,τ6,τ8,τ11,τ12}	101111010011	88.000000
{τ2,τ4,τ8,τ10,τ11,τ12}	010100010111	88.000000
{τ2,τ4,τ6,τ10,τ11,τ12}	010101000111	88.000000
{τ1,τ5,τ6,τ9,τ10,τ12}	100011001101	84.000000
{τ1,τ3,τ7,τ9,τ10,τ11,τ12}	101000101111	84.000000
{τ4,τ5,τ6,τ10,τ11,τ12}	000111000111	88.000000
{τ1,τ4,τ5,τ6,τ8,τ9,τ12}	100111011001	92.000000
{τ1,τ2,τ3,τ5,τ6,τ8,τ9,τ11}	111011011010	84.000000
{τ1,τ4,τ6,τ9,τ10,τ11}	100101001110	92.000000
{τ1,τ3,τ4,τ6,τ7,τ12}	101101100001	92.000000
{τ1,τ3,τ5,τ9,τ10,τ11}	101010001110	84.000000

You can repeat or iterate this experiment to a sufficiently appropriate number of times; however, we have repeated this experiment for 1000 generations to generate new test subsets. It was noticed that after 1000 generation, we did not find any change in the newly generated test subset, which has a better branch coverage than its predecessor generation. Hence, one of the best test subsets is chosen whose branch coverage is about 92%. The value of this subset was {1τ2, 1τ4, 1τ6, 1τ7, 1τ10, 1τ11}, and its coverage is 92. In this way, we can use a GA in test case prioritization.

In the next section, we discuss the application of partial swarm optimization (PSO) and differential evolution (DE) in test case generation for the modified code of the program.

10.3 TEST CASE GENERATION FOR MODIFIED CODE USING PSO-TVAC

The regression testing process involves several interesting research problems, such as how to find obsolete and redundant test cases, how to repair test cases, how to selectively and efficiently rerun test cases, how to create a new compelling test case,

TABLE 10.2
Selected Operation Based on a GA for a Triangle Problem

Test Subset	Binary Representation of Test Suite	Fitness (branch coverage of the subset)
{τ3,τ5,τ7,τ8,τ11}	001010110010	84.000000
{τ1,τ2,τ3,τ5,τ6,τ9,τ10,τ11}	111011001110	84.000000
{τ1,τ4,τ6,τ7,τ9}	100101101000	92.000000
{τ1,τ3,τ4,τ5,τ6,τ8,τ11,τ12}	101111010011	88.000000
{τ1,τ4,τ5,τ6,τ7,τ8,τ9,τ10,τ11}	100111111110	92.000000
{τ1,τ2, τ6,τ9,τ10,τ12}	110001001101	84.000000
{τ1,τ6,τ8,τ9,τ11}	100001011010	84.000000
{τ4,τ5,τ8,τ9,τ10,τ11,τ12}	000110011111	92.000000
{τ2,τ3,τ4,τ5,τ6,τ8,τ9,τ10,τ11}	011111011110	92.000000
{τ1,τ4,τ6,τ7,τ9}	100101101000	92.000000
{τ2,τ4,τ8,τ10,τ11,τ12}	010100010111	88.000000
{τ2,τ4,τ6,τ10,τ11,τ12}	010101000111	88.000000
{τ1,τ4,τ5,τ6,τ7,τ8,τ9,τ10,τ11}	100111111110	92.000000
{τ1,τ3,τ7,τ9,τ10,τ11,τ12}	101000101111	84.000000
{τ4,τ5,τ6,τ10,τ11,τ12}	000111000111	88.000000
{τ1,τ4,τ5,τ6,τ8,τ9,τ12}	100111011001	92.000000
{τ1,τ2,τ3,τ5,τ6,τ8,τ9,τ11}	111011011010	84.000000
{τ1,τ4,τ6,τ9,τ10,τ11}	100101001110	92.000000
{τ1,τ3,τ4,τ6,τ7,τ12}	101101100001	92.000000
{τ1,τ3,τ5,τ9,τ10,τ11}	101010001110	84.000000

and the basis for creating new test cases, and so on. These are interesting research problems, but here we are merely concerned with creating new practical test cases for modified code.

Whenever we request a set of new test cases after some modification in the program, the first analysis invokes the following questions:

- What will be the process followed to effectively generate new test cases?
- On what basis will a new test case generation process be followed?
- How will we justify whether the test cases generated have better code coverage?

We already discussed the NP-completeness of the test case generation process and we concluded that the optimization-based approach is best suited for generating near-optimal solutions.

What is PSO-TVAC?

PSO-TVAC is a variant of PSO in which particles velocity is modified (acceleration) according to the time, hence the name PSO with time-varying acceleration

TABLE 10.3
Crossover Operator

Test Subset	Binary Representation of Test Suite	Fitness (branch coverage of the subset)
{τ2,τ3,τ4,τ6,τ8,τ12}	011101010001	88.000000
{τ5,τ6,τ9,τ12}	000011001001	84.000000
{τ1,τ2,τ3,τ4,τ5,τ6,τ7,τ8,τ9,τ12}	111111111001	92.000000
{τ2,τ3,τ4,τ5,τ6,τ8,τ9,τ11}	011111011010	92.000000
{τ2,τ3,τ4,τ6,τ7,τ8,τ9,τ11,τ12}	011101111011	92.000000
{τ1,τ3,τ4,τ8}	101100010000	84.000000
{τ1,τ2,τ3,τ5,τ6,τ7,τ8,τ10}	111011110100	84.000000
{τ1,τ2,τ5,τ8,τ9,τ10}	110010011100	84.000000
{τ2,τ3,τ4,τ7,τ8,τ9,τ11,τ12}	011100111011	92.000000
{τ2,τ3,τ4,τ6,τ9,τ10,τ11}	011101001110	92.000000
{τ1,τ3,τ6,τ7,τ12}	101001100001	84.000000
{τ2,τ3,τ5,τ7,τ9}	011010101000	84.000000
{τ1,τ2,τ3,τ8,τ12}	111000010001	76.000000
{τ1,τ2,τ6,τ9}	110001001000	84.000000
{τ1,τ2,τ7}	110000100000	84.000000
{τ2,τ6,τ10}	010001000100	76.000000
{τ2,τ5,τ6,τ7,τ9,τ11}	010011101010	84.000000
{τ1,τ2,τ3,τ4,τ5,τ6,τ12}	111111000001	88.000000
{τ5,τ8,τ9,τ11,τ12}	000010011011	84.000000
{τ1,τ4,τ7,τ10,τ11}	100100100110	92.000000

(PSO-TVAC). This algorithm considers three random parameters (w, c_1, and c_2) whose value changes in each generation.

Why use a PSO-TVAC?

1. Simple algorithm, hence easy to implement
2. Domain undependability, suitable to use with most of the domains
3. Easy adaptability, making it easier to adapt by changing a few inherent parameters

Apart from the listed features, it also makes it an ideal candidate for new test case generation in regression testing.

10.3.1 Particle Initialization in PSO-TVAC

The way of working of PSO-TVAC is the same as we see in every swarm-based meta-heuristic technique; here also every particle represents a solution that moves in the whole search domain to find an optimal solution. It all depends on the way we guide our particles to reach at an optimal solution. Here, we first identify the section of the

TABLE 10.4
After-Mutation-Based Test Subset Generation by a Genetic Algorithm for Triangle Problem

Test Subset	Binary Representation of Test Suite	Fitness (branch coverage of the subset)
{τ1,τ2,τ8,τ10,τ11}	110000010110	64.000000
{τ1,τ5,τ8,τ9,τ10,τ11,τ12}	100010011111	84.000000
{τ1,τ2,τ3,τ4,τ6,τ7,τ9,τ10,τ11,τ12}	111101101111	92.000000
{ τ2,τ3,τ5,τ6,τ10,τ12}	011011000101	76.000000
{τ1,τ2,τ7,τ9,τ10}	110000101100	84.000000
{τ1,τ2,τ3,τ5,τ7,τ8,τ12}	111010110001	84.000000
{τ1,τ2,τ4,τ5,τ6,τ9,τ12}	110111001001	92.000000
{τ1,τ2,τ6,τ8,τ9,τ11}	110001011010	84.000000
{τ1,τ2,τ3,τ6,τ7,τ10,τ11,τ12}	111001100111	84.000000
{τ1,τ2,τ3,τ4,τ5,τ6,τ8,τ9,τ10}	111111011100	92.000000
{τ1,τ4,τ6,τ8,τ9}	100101011000	92.000000
{τ1,τ3,τ5,τ6}	101011000000	76.000000
{τ3,τ5,τ11,τ12}	001010000011	76.000000
{τ1,τ2,τ8,τ9,τ12}	110000011001	84.000000
{τ3,τ6,τ7,τ8,τ9,τ12}	001001111001	84.000000
{τ2,τ4,τ6,τ8,τ9,τ10,τ12}	010101011101	92.000000
{τ3,τ5,τ6}	001011000000	76.000000
{τ1,τ3,τ4,τ5,τ8,τ9,τ10,τ11,τ12}	101110011111	92.000000
{τ1,τ2,τ3,τ5,τ6,τ9}	111011001000	84.000000
{τ1,τ2,τ5,τ7,τ8,τ10,τ11,τ12}	110010110111	84.000000

code for the modification and accordingly we transverse the test cases. As far as the velocity of the particles is concerned, they are generated randomly from the input domain range. The particles that achieve the highest code coverage capability will act as the particles having the `gbest` position while those that are showing a better value within their own code coverage will act as particles having the `pbest` position.

10.3.2 Particle Update Rule

In every iteration, the position of the particles in a population is updated according to the fitness function, which we have already chosen according to the code coverage capability.

Let's take an example to understand how test cases are generated by the proposed algorithm. In Table 10.6, five modification-traversing test cases on delta version P' are given with their code coverage capability. Suppose the input domain range for input variables 'a' and 'b' is {1 to 50} and the maximum number of iterations are 2. Here `pbest` of a test case is its own achieved code coverage capability, that is, for test case T4, `pbest` is 64%, and `gbest` is the highest achieved code coverage capability among them, that is, 69%.

TABLE 10.5
Generation 1

Test Subset	Binary Representation of Test Suite	Fitness (branch coverage of the subset)
{τ1,τ2,τ8,τ10,τ11}	110000010110	64.000000
{τ1,τ5,τ8,τ9,τ10,τ11,τ12}	100010011111	84.000000
{τ1,τ2,τ3,τ4,τ6,τ7,τ9,τ10,τ11,τ12}	111101101111	92.000000
{ τ2,τ3,τ5,τ6,τ10,τ12}	011011000101	76.000000
{τ1,τ2,τ7,τ9,τ10}	110000101100	84.000000
{τ1,τ2,τ3,τ5,τ7,τ8,τ12}	111010110001	84.000000
{τ1,τ2,τ4,τ5,τ6,τ9,τ12}	110111001001	92.000000
{τ1,τ2,τ6,τ8,τ9,τ11}	110001011010	84.000000
{τ1,τ2,τ3,τ6,τ7,τ10,τ11,τ12}	111001100111	84.000000
{τ1,τ2,τ3,τ4,τ5,τ6,τ8,τ9,τ10}	111111011100	92.000000
{τ1,τ4,τ6,τ8,τ9}	100101011000	92.000000
{τ1,τ3,τ5,τ6}	101011000000	76.000000
{τ3,τ5,τ11,τ12}	001010000011	76.000000
{τ1,τ2,τ8,τ9,τ12}	110000011001	84.000000
{τ3,τ6,τ7,τ8,τ9,τ12}	001001111001	84.000000
{ τ2,τ4,τ6, τ8,τ9,τ10,τ12}	010101011101	92.000000
{ τ3,τ5,τ6}	001011000000	76.000000
{τ1,τ3,τ4,τ5,τ8,τ9,τ10,τ11,τ12}	101110011111	92.000000
{τ1,τ2,τ3,τ5,τ6,τ9}	111011001000	84.000000
{τ1,τ2,τ5,τ7,τ8,τ10,τ11,τ12}	110010110111	84.000000

TABLE 10.6
Test Case and Fitness

Test Cases	Fitness	pbest	gbest
{-1, 3}	64%	{-1, 3}	{6,2}
{2, 10}	64%	{2, 10}	
{6, 2}	69%	{6, 2}	
{2, 13}	64%	{2, 13}	
{7, -2}	69%	{7, -2}	

In the test case generation problem, there may be a situation in which more than one test case achieves the same highest value of code coverage capability. In that case, any test case among them can be selected as `gbest` of that iteration because `gbest` is used to guide the search to optimal solutions.

Suppose randomly generated initial velocity for each test case is {3,6}, {5,4} {15,17}{42,12}{24,35}, respectively.

Before updating the velocity and position of each test case, first we have to calculate the value of ω by using value of ω as

$$\omega = 0.75.$$

Here, the value of r_1 is 0.5 and r_2 is 2.5 while the value of ω is updated in each iteration. Suppose the values of r_1 and r_2 are 0.3 and 0.6, respectively, as these are the random numbers in the range [0, 1]. In this example, the calculated velocity and position of each test case is rounded as input variables have an integer data type.

Now, the velocity and position of each particle are updated by Equations 4.5 and 4.6 as such:

For test cases T1, T2, T3, T4, and T6, in first iteration, the velocity is updated as

$$v_1^1 = 0.75 * (3, 6) + 0.5 * 0.3(-1, 3) - (-1, 3) + 2.5 * 0.6(6, 2) - (-1, 3)$$

$$= \text{round}(12.75, 3)$$

$$= (13, 3).$$

Now the position of test case T1 is updated as

$$= (13, 3) + (-1, 3)$$

$$= \text{round}(12, 6)$$

$$= (12, 6).$$

Similarly, the velocity and position of T2, T3, T4, and T6 are calculated. The updated values are shown in Tables 10.7 and 10.8.

Updated Positions:

In Table 10.7, V_i is the randomly generated initial velocity of the test case while V_1 is the updated velocity in the first iteration. Similarly, in Table 10.8, X_i is the initial position of the test case while X_1 is the updated position in first iteration. The fitness is calculated as Fitness (X_1) for each updated position using Equation 10.1. The highest own fitness value of a test case between initial and first iteration is selected as `pbest`, and the test case that has highest value of fitness among `pbest` is taken as `gbest`.

TABLE 10.7

Updated Velocity of Particles

V_i	V_1
{3, 6}	{13, 3}
{5, 4}	{10, -9}
{15, 17}	{12, 13}
{42, 12}	{37, -7}
{24, 35}	{16, 20}

TABLE 10.8
Updated Positions of Test Cases

X_i	Fitness X_i	X_i	Fitness X_i	pbest	gbest
{-1,3}	64%	{12,6}	69%	{12,6}	{39,6}
{2,10}	64%	{12,1}	69%	{12,1}	{6,2}
69%	{18,15}	49%	{6,2}	{2,3}	64%
{39,6}	70%	{39,6}	{7,-2}	69%	{23,18}
48%	{7,-2}				

Similarly in the second iteration, the value of ω is updated as

$$\omega = 0.4.$$

Now the velocities and positions of each test case are updated, shown in Tables 10.9 and 10.10.

Updated Velocities:

TABLE 10.9
Updated Velocities of Test Cases

V_i	V_1	V_2
{3,6}	{13,3}	{46,1}
{5,4}	{10,-9}	{44,4}
{15,17}	{12,13}	{34,-10}
{42,12}	{37,-7}	{15,-3}
{24,35}	{16,20}	{28,-13}

Updated Positions:

Here, pbest contains the generated test cases that have higher code coverage capability as compared to the initial test cases. This example illustrates the process of

TABLE 10.10
Updated Test Cases

X_1	Fitness X_1	X_1	Fitness X_1	X_2	Fitness X_2	pbest	gbest
{-1,3}	64%	{12,6}	69%	{58,7}	74%	{58,7}	{58,7}
{2,10}	64%	{12,1}	69%	{56,5}	70%	{56,5}	{6,2}
69%	{18,15}	49%	{52,5}	64%	{6,2}	{2,13}	64%
{39,6}	70%	{54,3}	69%	{39,6}	{7,-2}	69%	{23,18}
48%	{51,5}	69%	{51,5}				

test case generation up to two iterations. The same process is repeated for the greater number of iterations.

10.4 TEST CASE GENERATION FOR MODIFIED CODE USING DE

Like PSO, DE can also be used in test case generation. DE is a stochastic direct search method. The choice of DE parameters F, Cr, and NP can have a large impact on optimization performance.

10.4.1 Population Initialization

DE is a population-based algorithm, and it works like GAs using similar operators: crossover, mutation, and selection. Population initialization in DE is done by generating a set of solution vectors within the given bounds by generating the random values between the lower bound $X_{j,min}$ and upper bound $X_{j,max}$ for each dimension of the solution vector, where *NP* is the size of initial population and *dimension* refers to the number of variables in each vector.

$$\vec{X}_{j,i,0} = X_{j,min} + rand_{i,j}\left[0,1\right]*\left(X_{j,max} - X_{j,min}\right), \tag{10.1}$$

where $1 \le i \le NP$,
$1 \le j \le$ Dimension, and
$0 < rand_{i,j}\ [0, 1] < 1$.

Fitness will be determined for each solution by measuring the code coverage capability value using a code coverage analysis tool, like Gcov and others.

10.4.2 Update Rule

In each iteration of the proposed algorithm, each solution/test case has to pass through the mutation and crossover operators, and then the position of each particle/test case is updated according to the fitness function, that is, code coverage capability, so that it can reach to a global optimal solution.

Let us take the same example that we used for PSO-TVAC to clearly describe the process of DE. In Table 10.11, five modification-traversing test cases on delta version

TABLE 10.11
Initial Test Cases

Test Cases	Fitness
{-1, 3}	64%
{2, 10}	64%
{6, 2}	69%
{2, 13}	64%
{7, -2}	69%

P' are given with their code coverage capability. Let the input domain for input variable a, and b is {1 to 50} and the maximum number of iterations is 2. These initial vectors are known as target vectors.

10.4.3 Mutation

During mutation, for each target vector, a donor vector is created by randomly choosing three initial vectors from the initial pool of target vectors. First of all, a vector known as the target vector is chosen from the current generation and is added to the scaled difference of two other vectors, where `r1, r2, and r3` are the positions of the vectors and `i` is the position of the mutated vector.

For example, if we want to create a mutation vector for the first solution, we will randomly select three vectors from the pool of five solutions. If the positions of those vectors are 2, 4, and 5, then vector {2, 10} will work as $\vec{X}^{1}_{2,0,}$ {2, 13} will be $\vec{X}^{1}_{4,0}$ and {7, −2} will be $\vec{X}^{1}_{5,0}$.

Now with the help of Equation 10.2, we will identify the value of donor vector for first solution.

$$\vec{V}_{i,G} = \vec{X}_{r1^i,G} + F * \left(\vec{X}_{r2^i,G} - \vec{X}_{r3^i,G} \right), \tag{10.2}$$

where $\vec{V}_{i,G}$ is the donor vector.

$\vec{X}_{r1^i,G}$, $\vec{X}_{r2^i,G}$ and $\vec{X}_{r3^i,G}$ are three chosen vector from current generation G.

F is scalar number [0.4–1.0].

$r_1 \neq r_2 \neq r_3$ are mutually exclusive.

So for the first solution, the following donor vector will be generated:

$$\vec{V}_{1,G} = \vec{X}_{2^1,G} + F * \left(\vec{X}_{4^1,G} - \vec{X}_{5^1,G} \right)$$

= {2, 10} + 0.2 * ({2, 13}−{7, −2})
= {2, 10} + {−1, 3}
= {1, 13}

We will repeat this process for all five solutions. The updated values are shown in Table 10.12.

TABLE 10.12
Updated Test Cases after Mutation

Test Cases	Donor Vectors
{-1, 3}	{1,13}
{2, 10}	{6,-5}
{6, 2}	{2,15}
{2, 13}	{3,11}
{7, -2}	{-2,10}

10.4.4 Crossover

In order to enhance diversity within the population, crossing of donor vector $\vec{V}_{i,G}$ with target vector $\vec{X}_{i,G}$ is done, where their components are exchanged to form trial vector $\vec{U}_{i,G}$ by using Equation 10.3:

$$\vec{U}_{j,i,G} = \vec{V}_{j,i,G} \quad if\ \left(rand_{i,j}\left[0,1\right] \leq Cr\ or\ j = j_{rand}\right) \tag{10.3}$$

$$\vec{X}_{j,i,G}\text{, otherwise,}$$

where $0 < rand_{i,j}[0, 1] < 1$,
Cr is the crossover rate, and
j_{rand} ε [1, 2, . . . , D].

For example, to create the trial vector for the first solution, we will take first donor vector from the set of donor vectors, that is, {1, 13}, and combine it with the corresponding target vector, that is, {−1, 3}, since in testing, the crossover rate is taken as low because we do not need much diversity. We will choose the value of *Cr* accordingly. Let the value of *Cr* is 0.4. Now for each dimension, we will generate the random values between 0 to 1. If the generated values are (0.1, 0.5), then the generated trial vector will be having the first component of donor vector and the second component will be of target vector. When this process is applied for all the vectors, these trial vectors are generated. The updated values for the test cases after mutation and crossover operators are shown in Table 10.13.

10.4.5 Selection Operator

Here, the survival of the fittest principle is implemented to direct the winner of the race for the next generation. The fitness value of trial vector is generated, and then the fitness of the target vector and the trial vector are compared, and the better one is chosen. Since the objective of the problem is to generate a vector that has a higher value of fitness, the test case with a better fitness is always chosen. For this problem, the selected candidate will be chosen by Equation 10.4:

TABLE 10.13
Updated Test Cases after Mutation and Crossover

Test Cases	Donor Vectors	Trial Vector
{-1, 3}	{1,13}	{1,3}
{2, 10}	{6,-5}	{6,-5}
{6, 2}	{2,15}	{6,15}
{2, 13}	{3,11}	{3,13}
{7, -2}	{-2,10}	{7,10}

TABLE 10.14
Final Test Cases

Test Cases	Fitness	Trial Vector	Fitness	Final Vector
{-1, 3}	64%	{1,3}	65%	{1,3}
{2, 10}	64%	{6,-5}	61%	{2, 10}
{6, 2}	69%	{6,15}	73%	{6,15}
{2, 13}	64%	{3,13}	70%	{3,13}
{7, -2}	69%	{7,10}	53%	{7, -2}

TABLE 10.15
Updated Test Cases after Mutation of Second Generation

Test Cases	Fitness	Donor Vector
{1,3}	65%	{3,-7}
{2,10}	64%	{-2,11}
{6,15}	73%	{-4,13}
{3,13}	70%	{1,5}
{7,-2}	69%	{4,3}

TABLE 10.16
Updated Test Cases after Mutation and Crossover

Test Cases	Donor Vectors	Trial Vector
{1,3}	{3,-7}	{1,-7}
{2,10}	{-2,11}	{-2,10}
{6,15}	{-4,13}	{-4,13}
{3,13}	{1,5}	{1,13}
{7,-2}	{4,3}	{7,3}

$$\vec{X}_{i,G+1} = \vec{U}_{i,G} \quad if\ f\left(\vec{U}_{i,G}\right) \geq f\left(\vec{X}_{i,G}\right) \tag{10.4}$$

$$\vec{X}_{i,G}\ if\ f\left(\vec{U}_{i,G}\right) < f\left(\vec{X}_{i,G}\right),$$

where $f(X)$ is the fitness function to be maximized.

Similarly, in the second iteration, the first mutation will be applied, and the donor vector is generated, as shown in Table 10.15.

Then the crossover is applied, and trial vector is generated as shown in Table 10.16.

TABLE 10.17
Final Test Cases after the Second Generation

Test Cases	Fitness	Trial Vector	Fitness	Final Vector
{1,3}	65%	{1,-7}	72%	{1,-7}
{2,10}	64%	{-2,10}	60%	{2, 10}
{6,15}	73%	{-4,13}	73%	{-4,13}
{3,13}	70%	{1,13}	75%	{1,13}
{7,-2}	69%	{7,3}	59%	{7, -2}

Finally, the selection operator is applied to generate the next generation. The results are tabulated in Table 10.17.

This example illustrates the process of test case generation up to two iterations. The same process is repeated for the greater predefined number of iterations.

11 Application of Genetic Algorithms and Partial Swarm Optimization in Cloud Computing

11.1 INTRODUCTION

We looked at how a nature-inspired algorithm can be used to generate software test cases in the last chapter. Another key use of optimization algorithms in the computer science sector is discussed in this chapter. This issue has something to do with cloud computing, which is a relatively new technology. While many optimization problems can be framed in the cloud computing arena, this chapter focuses on workflow scheduling challenges in the cloud. We begin with discussing cloud computing technology before moving on to cloud computing problems in which optimization approaches can be used. Finally, we target workflow scheduling problems and discuss various steps required to apply optimization algorithms. We also discuss essential tools and simulators that are necessary to implement an optimization algorithm.

11.2 CLOUD COMPUTING

Cloud computing is a business model that uses parallel computing distributed computing, grid computing, and virtualization technologies to provide online services to users. Its purpose is to provide ubiquitous, convenient, on-demand network access to a shared pool of configurable computing resources (e.g., networks, servers, storage, applications, and services) that can be rapidly provisioned and released with minimal management effort.

11.2.1 Types of Cloud

Depending on the type of service required, the cloud can be divided into three categories based on the type of services shared for application implementation: Infrastructure as a Service (IaaS), Platform as a Service (PaaS), and Software as a Service (SaaS).

11.2.1.1 IaaS

IaaS is a method of provisioning infrastructure resources on demand and at a low cost. These resources could include hardware, storage, servers, data center space, and network components such as software. These resources can be used to perform any computation. This technique is known as IaaS because these resources are part of the

DOI: 10.1201/9781003313649-11

infrastructure. Many well-known companies own these resources and are referred to as cloud providers. Amazon EC2, GoGrid, and Flexiscale are some examples.

11.2.1.2 PaaS

Some businesses make their computing platform available for use. Operating system support and software development frameworks are among the platform resources available. Companies that provide platform-related services include Google App Engine, Microsoft Azure, and Force.com to name a few.

11.2.1.3 SaaS

SaaS clouds are used to provide software-related services to clients, and they are implemented by giving on-demand applications via the internet. Salesforce.com, Rackspace, and SAP Business by design are a few examples of companies that support these services.

11.3 WORKFLOW AND ITS SCHEDULING

The execution of a highly complex application on a cloud is considered a workflow scheduling problem. The application can be broken down into several tasks that may be dependent or independent to each other, and they may also be arranged in some order. Workflow is a collection of tasks ready to be executed on a set of virtual machines (VMs). Consider printing a document from the cloud. Printing a document can be divided into several steps. The first step will be to choose an editor and write an application. The second task is to select a printer from the list provided, and the third task is to print the page. As you can see, these tasks are all interconnected and must be completed in the correct order. A complex application necessitates a large number of tasks, some of which may be highly complex. Workflows are created and used in a wide range of engineering domains. Scientific workflows are those used in cloud/grid computing to solve scientific problems.

A cloud is made up of resources where workflow can be executed to complete a specific application. A graph can be used to display various cloud resources. It can be drawn to show the storage and processing capacity of VMs and the cost of data transfer between virtual machines. The graph in Figure 11.1 represents a cloud that contains three virtual machines' (VM_1, VM_2, VM_3) storage capacity (S_1, S_2, S_3) and data transfer cost as (D12, D23, D13).

For processing any application on a cloud, the workflow has to be prepared, and then a scheduler is used to map this workflow on cloud resources. Let us discuss an example of a sample workflow. This workflow contains five tasks and should be executed on a cloud having three VMs.

An example of workflow schedule is shown in Table 11.1.

11.3.1 How to Choose an Appropriate Cloud Provider for Running Workflows

With so many cloud service providers on the market, the user's choice of a cloud provider is based on the price offered by the provider for giving services. A user will

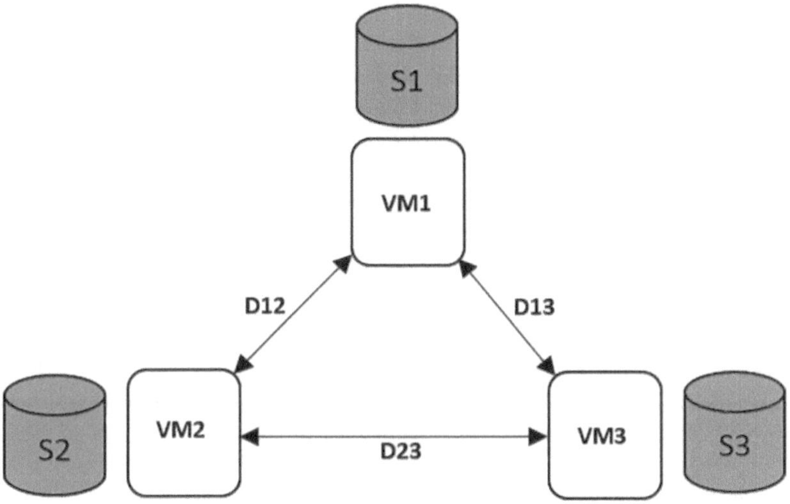

FIGURE 11.1 Graphical representation of cloud.

TABLE 11.1
Example Workflow Schedule

Tasks	t1	t2	t3	t4	t5
Computing Resources	VM1	VM2	VM1	VM2	VM3

prefer a provider whose services are less expensive. Therefore, to continue in business, cloud service providers must lower the cost of their services while maintaining service quality. The cloud service provider does this by fixing a few goals for running user workflows. Some cloud providers are just interested in providing the consumer with the cheapest available solution for their workflow. Some cloud providers may offer a low cost for user workflows by setting an equal load on each VM. Some cloud providers strive to maximize their profit by conducting workflow so that it spends the least amount of time on their resources and costs the least amount of money.

The complexity of workflow scheduling emerges when the number of VMs is large enough to apply classic methods like FCFS (first come, first serve), round-robin, and SJF (shortest job first) to identify the lowest cost or time. It's challenging to reduce the computational cost of such workflows, and establishing more cost-effective scheduling policies involves a thorough review of current job scheduling methods. Optimization methods may help with such issues. They can be used to design the most efficient timetable for the least amount of money.

Consider how an optimization algorithm can be used to reduce the cost of workflow scheduling. An optimization algorithm requires a mapping that defines how a workflow schedule can be represented as a solution. Once the mapping is determined, a few workflow schedules will generate the initial population for the optimization algorithm. These solutions will be updated to create new workflow schedules using the fitness function until the optimal workflow schedule is developed. A fitness function that adds execution and transfer costs to calculate the cost of executing workflow can generate this schedule. A new fitness function must be created if load balancing is required in addition to cost minimization. When the same VMs are used to execute multiple workflow tasks, a penalty value may be added into this modified fitness function. In this modified fitness function, a penalty value may be added when the same VMs are used to execute multiple workflow tasks. This penalty will be imposed to prevent the use of workflow schedules that do not utilize all VMs.

11.3.2 Workflow Scheduling Problem Formulation

The determination of an effective workflow schedule is a very complex problem that necessitates an optimization algorithm. The effective scheduling policy assists us in minimizing total execution time and cost, as well as load balancing on all computing resources. Two major factors influence workflow costs. The first factor is related to execution cost, associated with processing units used in the cloud. Several processing units (referred to as VMs) are involved in the cloud computing environment. These processors used in the cloud are heterogeneous and have varying processing capabilities. The processing ability varies and is dependent on the number of memory units and the computing capacity of the machine. Each VM has a different execution cost. In addition to the execution cost, the transfer cost also plays an essential role in defining the overall cost of the workflow. During the execution of tasks, some tasks are dependent on the output of other tasks, so the transfer of data takes place between these tasks. Therefore, transferring data from one task to another is treated as a communication cost in the paper.

A directed cyclic graph (DAG) can be used to display details of a workflow schedule, such as which VMs are used to run the task and whether or not data transfer is required. A workflow application is mapped using a DAG, which is represented by 1 (b). In this case, the graph is represented by G = (V, E). The set of tasks is denoted as T = T1, T2, ... , Tn, and the data dependency between the tasks is represented as E, that is, fp, c = (tp, tc) E, where tp represents the data produced and consumed by `tc`.

We have taken a set of tasks

T = {1, 2, . . ., i},

a set of *VMs = {1, 2, . . ., j}*,

and a set of storage units *S = {1, 2, ... , k}*.

It is assumed that a task's average computation time is *Ti* on a computing unit *VMj* for a certain known input size. Also, the unit data access cost is assumed as *Di, j*. Here $D_{i,j}$ represents the cost is already known from VM_i to VM_j and $D_{i,j} = D_{j,i}$ *for all i, j* ∈ *N*, where *N* represents the nodes.

Bandwidth between the processing units decides the communication cost. The service provider decides the subscription cost of the resources. Also, the communication cost of transferring data between the VMs has been charged on a per-second basis. Here, the objective is to minimize the total cost of computation by optimally assigning the workflow task.

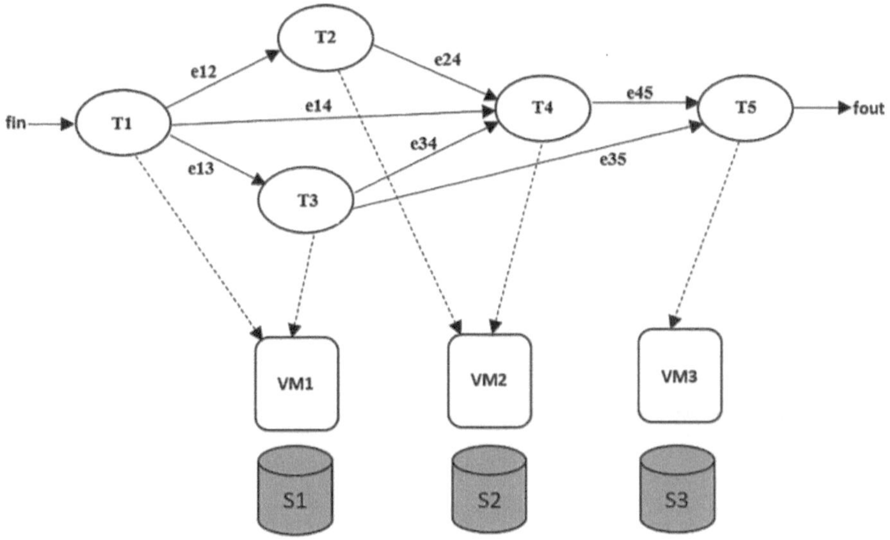

FIGURE 11.2 An example of a workflow model with VM and storage.

Figure 11.2 shows the workflow model with five nodes, and each node represents an individual task, which includes several instructions. Arrows are used to represents the dependencies between the tasks. In the workflow model, task *T1* represents the root that takes the input file, and *T5* is the last task that gives the output file. For example, *T1* produces output after completing *e12* (*T1, e12, T2*), the same concept works for all tasks such as (*T1, e13, T3*), (*T1, e14, T4*), (*T2, e24, T4*), (*T3, e35, T5*), and so on. Here, the first part represents the starting task, the middle part is used to denote the edge (output), and the last part shows the intermediate task or the end task that takes the output from the previous task as input. The edge-weight (*ei1, i2*) represents the numeric values between two tasks; that is, $i1 \in T$ and $i2 \in T$. The figure shows the three VMs interconnected to each other with varying bandwidths. Each VM has its storage unit that has been represented as *S1, S2,* and *S3*, respectively. Here, the objective is to minimize the total cost of computation by assigning the workflow task in an optimal way.

Since the cloud provider wants to attract more and more users by offering them the minimum cost to execute workflows. They provide a scheduler that can generate an optimal schedule for user workflow.

11.3.3 Generating Optimal Workflow with the Help of Partial Swarm Optimization

Designing an optimal scheduler is not easy. An optimal schedule cannot be generated by mapping tasks to any random VMs. It can be created with the help of an optimization algorithm that defines an optimal strategy that can identify which task should be executed on which virtual VM. A randomized algorithm can be

TABLE 11.2
Workflow Representation in Tabular Form

t1-VM1	t2-VM2	t3-VM1	t4-VM2	t5-VM3

used to design such a scheduler. For implementing the optimization algorithm, we must provide two things: First is how the workflow schedule should be represented, and second is how the cost should be calculated. The workflow schedule will be expressed to apply the algorithm, and fitness will be required to identify the best solution. Previously we have seen how we can define a workflow on a cloud application; see Table 11.2.

After representing the workflow, we need to calculate the fitness of the workflow. The following method can be used to calculate the fitness of workflow.

11.3.4 Fitness Function for Workflow Scheduling

To find the best workflow schedule, an optimization algorithm needs a suitable fitness function. The fitness function is essential in determining the best workflow schedule for the given problem. To solve the problem, a variety of fitness functions can be designed. In the given situation, we want to minimize workflow cost, which is determined by the execution and transfer (communication) costs. In addition to reducing workflow costs, it should be reflected in the fitness function if you want to add more constraints. If we also want to perform load balancing by utilizing all VMs, we have to add some penalty factor in the fitness function in addition to the normal cost. Since the cloud computing environment consists of heterogeneous VMs, the execution cost of the VMs also varies. In addition to the execution cost, the transfer cost is also added. Generally, we use a fitness function in which the schedule cost will be identified with the help of execution cost and communication cost. However, in exceptional cases, like load balancing and assigning more priorities to some workflow, we need to include additional information in the fitness function. For example, we can emphasize those schedules in which dependent tasks are executed on the same virtual machine. This kind of assignment will minimize the transfer cost, which will help us produce an optimal workflow as early as possible.

Moreover, for proper utilization of all resources, we can add a penalty function in the fitness function. This fitness function will produce more cost if fewer virtual machines are utilized. The number of unused VMs will be added to the penalty value.

Depending on the type of problem, we want to handle our fitness function requires these three components. It constitutes the cost of using the computing power in millions of instructions per second (MIPSs) multiplied by the cost per unit. The execution cost is computed using Equation 11.1.

Execution cost:

$$C_{exe} = \sum_{t=1}^{k} \left(\frac{It * CpTi}{MIPSi} \right), \tag{11.1}$$

where
VMi = VM on which task *t* is executed,
It = number of instructions in task *t*,
MIPSi = MIPS value of the VM executing task *t*, and
CpTi = cost per unit of time execution of the VM executing task *t*.

The transfer cost is calculated with the help of the dependency graph. Communication costs will be added if parent task and child tasks are at different VMs. Let all the virtual machines {*VM1, VM2, ... , VMn*} have taken as nodes {*V1, V2, ... , Vn*} of graph *G*. The communication cost (computed by Equation 11.2) of transferring data from *VM1* to *VMn* is the weight of edge *E {V1, Vn}*.

Communication cost:

$$C_{comm} = \sum E\{Vi, Vj\} \tag{11.2}$$

For each communication between *i* and *j*, the penalty is calculated on the basis of the number of unused *VMs*. We have assumed a factor of 1000 for each unused *VM*.

For load-balancing purpose:

$$Penalty(P) = \left(\prod_{i=1}^{n} Ni * No.\,of\,VMs\,unused * 1000 \right), \tag{11.3}$$

where *Ni* = number of tasks executed on VM.

So, the total cost would be the sum of three factors: the execution cost, the communication cost, and the penalty.

$$Computation\,Cost\left(C_{total}\right) = C_{exe} + C_{comm} + P \tag{11.4}$$

So, the final objective function is to minimize (C_{total}).

11.4 APPLICATION OF PARTIAL SWARM OPTIMIZATION FOR GENERATING OPTIMAL WORKFLOW SCHEDULE

In this section, we explain the application of the partial swarm optimization (PSO) algorithm for generating an optimal schedule for a given workflow. As discussed in previous chapters, PSO is a population-based algorithm in which solutions should be represented by using some encoding. In this problem, the particle will be denoted by random schedules, and fitness represents the cost of those schedules.

If load balancing is not required, then each workflow schedule's cost can be calculated by the task execution cost and the communication cost.

So, for each random workflow, we check the execution cost of each task and add it to the communication cost if the data transfer is required between two tasks.

As an example, let us take one workflow and see how the cost will be calculated. This cloud contains three heterogeneous VMs with varying configurations and costs.

This is a random mapping that a particle will represent. From the graph, all tasks will be executed on the three VMs. However, a parallel execution of the task is not possible because some tasks are interdependent. From the figure, first task t1 is the main task, and once it is executed, only other tasks can be completed after execution t1 will transfer its data to other dependent tasks, t2 and t3. Tasks t2 and t3 can be achieved in parallel. After executing t2 and t3 will share its data with t4, then t4 can be executed. After t3 and t4, t5 can be executed.

In the given example:

- Schedule Generated: t1vm1 t2vm2 t3vm1 t4vm2 t5vm3.
- Cost Computation: The execution cost will be the running cost of t1 on vm1, t2 on vm2 and similarly other execution costs can be calculated. The communication costs will also be calculated. In the given graph, the data transfer is shown by edges. Since the data transfer cost will be provided by the cloud provider. These values can be multiplied by the weights of edges to calculate communication costs.

For generating an optimal schedule for the application, we need to initialize the population and design the operations accordingly.

Optimal schedule PSO (random schedule, fitness function):

- Optimal Schedule Generation: For generating the optimal schedule using the PSO algorithm, we follow these steps:

Step 1: Generate the population size and dimension; initialize the iteration as 0 and max iterations (n).

Step 2: For a set of random workflows for representing particle positions, define `gbest` and `pbest`.

Step 3: Update the velocity and the position of each particle by following equations used in PSO: $V_i = \omega * V_i + c1 * r_1 * (P_i - X_i) + c2 * r_2 * (G - X_i) \ldots$ (11.6)

$$Xi = Xi + Vi \ldots, \tag{11.7}$$

where Xi is ith particle and V_i velocity of ith particle and P_i is the `pbest` of ith particle and G is the `gbest`.

Step 4: Update `pbest` and `gbest`.

Step 5: Update the iteration counter and check whether the max iteration is reached or the optimal schedule generated; if yes, terminate the algorithm; if not, repeat steps 2 through 4 with the updated positions and velocities.

11.4.1 An Example Application

Step 1: Let us take an example and see how PSO can be used in generating an optimal schedule. This workflow is composed of five tasks and will run on a cloud having four VMs.

First, we define two variables: One is the number of tasks, and the other is the number of VMs.

Number of tasks = 5
Number of VMs = 4
Max `itr` = 100(Maximum iterations)

After fixing the number of tasks and VMs, we use an encoding to represent the particles in PSO. In our example, we initialize five particles in the initial population.

#Particles = 4

For generating each particle, we define the dimension of the particle. In our example, the mapping requires five tasks on four VMs; therefore, we randomly initialize five variables of each particle in the range [0, 3]. We only choose integer values for initializing particles. Each assignment index of variable represents the task number, and the value within the index will represent the VM number. Let us take an example of such an assignment in Table 11.3.

This mapping defines that task 1 will be executed on the second VM; task 2 is running on the third VM, task 3 is running on the first VM, task 4 is running on the third VM, and task 5 is running on the fourth VM. After generating all the particles, we define the fitness of each particle.

This particle will be represented by 1 2 0 2 3.

Step 2: Generating the initial population

In PSO, the first step is to initialize the population. This population includes the initialization of both position and velocity vectors. We randomly generate four positions and velocities for the four particles. Here are the values of the positions and their corresponding finesses.

Remember only the positions of particles will be evaluated to judge the fitness of the schedule.

The fitness calculation is done with the help of the execution cost and the transfer cost. In the example data, these costs will be calculated by the following information.

Data required for calculating the execution cost and the transfer cost.

Data used for execution (Cost) calculation

vm_cost = {8, 9, 5, 6}; //Execution cost for the VM // Processing cost

vm_mips = {4, 5, 3, 6}; // Processing capacity

task_mi = {8, 10, 6, 12, 24}; //Size of Task in MI (million instruction)

task_transfer = Data transfer in MB from one task to task

{{0, 0.17, 0.20, 0.20, 0.21}, {0.17, 0, 0.20, 0.20, 0.21}, {0.20, 0.20, 0, 0.17, 0.22}, {0.20, 0.20, 0.17, 0, 0.22}, {0.21, 0.21, 0.22, 0.22, 0}, {0.21, 0.21, 0.22, 0.22, 0.17}, {0.18, 0.18, 0.19, 0.19, 0.20}, {0.18, 0.18, 0.19, 0.19, 0.20}}

TABLE 11.3
Example of Task Assignment

Index	1	2	3	4	5
VM number	1	2	0	2	3

TABLE 11.4
Positions of the Particle and Its Cost

Particles (X)	Cost: f(X)
2 1 1 0 2	109.48(Execution cost 106.13+Transfer Cost 3.34)
1 2 0 2 3	91.01(Execution cost 87.06+Transfer Cost 3.95)
3 1 1 0 1	107.42(Execution cost 104.00+Transfer Cost 3.42)
2 0 2 1 3	93.46(Execution cost 88.93+Transfer Cost 4.53)

Int [][] tran_vm_cost = VM transfer Cost for transferring file/data as MB/sec from one VM to other

{{2,3,4,1}, {5,4,3,6}, {6,8,7,9}, {3,5,2,7}};

Calculate Execution Cost as follows:

Execution Cost = (Size of Task in MI)/ (Processing capacity of VM) * Processing cost of VM;

Data Transfer Cost = (Size of data transfer between task) * VM Transfer Cost;

Defining gbest and pbest:
The initial population and its fitness cost are listed in Table 11.4.

After calculating the fitness of these particles, we need to find `gbest`. At this time, particle 1 2 0 2 3 is producing minimum fitness; therefore, it will work as `gbest` of the problem. Similarly, `pbest` values will be calculated. Since this is the first iteration of the algorithm, there will be no history of the movement. Therefore, the value of `pbest` and the position of particle will be same. Initial velocities V of particles X are shown in Table 11.5.

Step 3: Generating new velocities and position of particles and clamping
The values of `gbest` and `pbest` are used to generate new particles. Given that this is the first generation, the value of `pbest` will be equal to the particle position and `gbest` that we have already calculated. We can now calculate the new velocities based on these values. Because float values aren't allowed, we use the round function to convert them to integers. Furthermore, we use clamping if the velocity value exceeds a specific limit. As the results in Tables 11.6 and 11.7 show, new speeds and positions will be established.

TABLE 11.5
Initial Velocities of the Particles

Particles (X)	Velocities V
2 1 1 0 2	0 0 2 3 0
1 2 0 2 3	3 1 3 0 1
3 1 1 0 1	2 1 0 1 1
2 0 2 1 3	3 3 3 3 3

Step 4: Updating gbest and pbest
Now we update the new gbest and pbest and use these values for defining new velocities and positions of the particles. From the data, it is clear that 1 1 0 1 3 now has the minimum fitness. So this will act as gbest for new iterations. Similarly, the values of pbest can be decided accordingly.

Compare the fitness of old population X and new population Xnew, and keep the best in pbest. The new pbest is shown in Table 11.8.

New values for pbest and gbest will be used to define new velocities and positions. This process will be repeated again and again until the number of iterations become equals to the maximum number of iterations.

The final value of the optimal solution will be printed.

Step 5: In step 5, we update the iteration counter. Now this becomes 1, and the maximum iterations are 100. It has not reached to maximum value. We repeat steps 2 to 4.

TABLE 11.6
New Velocities of the Particles

New particle velocities V
0 0 2 3 0
3 1 3 0 1
2 1 0 1 1
3 3 3 3 3

TABLE 11.7
Updated Position and Its Cost Values of the Particles

Updated Position (X)	Cost: f(X)
2 1 3 1 3	82.65
3 0 3 0 3	76.98
0 3 2 1 0	103.68
1 1 0 1 3	80.77

TABLE 11.8
New pbest Values

New pbest
2 1 3 1 3
3 0 3 0 3
0 3 2 1 0
1 1 0 1 3

TABLE 11.9
Output Values on Simulator

Cloudlet ID	STATUS	Data Center ID	VM ID	Execution Time	Start Time	Finish Time
0	SUCCESS	2	3	10.02	0.1	10.12
1	SUCCESS	2	0	10.13	0.1	10.23
2	SUCCESS	2	1	10.13	0.1	10.23
3	SUCCESS	2	3	10	10.12	20.12
4	SUCCESS	2	3	10	20.12	30.12

This entire procedure can be implemented using the cloudsim framework. This framework provides the necessary support for simulating the cloud environment for programming; interested readers may refer to material available on the internet for further study. Now I show how the output is generated on the cloudsim framework. This is a sample output generated by the cloudsim framework. The outputs of the simulator are tabulated in Table 11.9.

We have discussed how the PSO algorithm can generate an optimal workflow schedule in a cloud computing environment. Similar to PSO, other randomized algorithms can also be used to generate optimal workflow.

11.5 APPLICATION OF A BINARY GENETIC ALGORITHM IN WORKFLOW SCHEDULING PROBLEM

We've seen how a real-parameter optimization algorithm can be used to solve workflow scheduling issues. Is it possible to solve this problem with a binary encoding algorithm?

Is it possible to represent a workflow schedule with a binary solution? Sometimes it is not possible to describe the solution directly in binary form. However, some modifications can be made to define the solution in binary. Remember, we are the ones who are representing the representation. In the following section, we will demonstrate how a binary genetic algorithm can solve this problem. Only encoding and fitness calculation will be discussed; the rest can be implemented by understanding how algorithms work.

11.5.1 Binary Encoding Used for Representing the Workflow Schedule

In real-parameter encoding, an array was used to represent the schedule in which the index represented the task number and the number of VMs assigned randomly in the problem. We can use a similar kind of representation only with some changes. In this representation, index values will be used to define the task number, and a set of consecutive bits can be used to represent the virtual machine number. Let us retake the same problem which we have taken in PSO. We want to map five tasks onto four machines. To identify a virtual machine uniquely, we need two binary bits. We generate an array of five bits and use the following method to identify the mapping as shown in Table 11.10.

TABLE 11.10
Workflow Schedule (task VM mapping)

Index	1	2	3	4	5
VM number	0	1	0	0	1

To generate the required mapping, we can combine two consecutive bits and define the number of VMs. In our mapping, 1 bit comes from the index value, and the other comes from the following index. If it is the last index, take the first bit. So this representation will define the schedule in the following way:

The first task will be mapped onto 01 VM, that is, the second VM.
The second task will be mapped onto 10 VM, that is, the third VM.
The third task will be mapped onto 00 VM, that is, the first VM.
The fourth task will be mapped onto 01 VM, that is, the second VM.
Finally, the last task will be mapped onto 10 VM, that is, the third VM.

In this way, we can identify the schedule and can calculate the fitness of the solution.

So, the workflow schedule for following solution will be as follows:

t1VM1	t2VM2	t3VM0	t4VM1	t5VM2

Calculating the fitness value of the solution: Once the schedule is identified, we can use the same fitness function as we used in PSO for defining the fitness of the solution.

Producing the new population: After generating the initial population and calculating the fitness, we can use traditional selection, crossover, and mutation operators again and again until either the optimal solution is developed or the maximum iterations are reached.

In addition to the real-parameter and binary encoding techniques, other encoding techniques may be used to solve the problem. Keep in mind that a better encoding strategy may affect the algorithm's performance. It can generate a perfect solution in far fewer iterations.

Conclusion: According to the no-free-lunch theory, no single algorithm is superior for solving all optimization problems. People are developing new algorithms to provide more and more efficient solutions to problems. As previously discussed, when solving problems with optimization algorithms, the representation of the solution and the design of the fitness function can improve the algorithm's overall performance. In the cloud computing environment, we can apply optimization algorithms to a variety of other problems. People are still looking for new ways to use nature-inspired algorithms.

References and Further Reading

[1] Yang, Xin-She, ed. *Nature-Inspired Algorithms and Applied Optimization*. Vol. 744. Springer, New York City, NY, 2017.

[2] Yang, Xin-She. *Nature-Inspired Optimization Algorithms*. Academic Press, Cambridge, MA, 2020.

[3] Chiong, Raymond, ed. *Nature-Inspired Algorithms for Optimisation*. Vol. 193. Springer, New York City, NY, 2009.

[4] Bozorg-Haddad, Omid, ed. *Advanced Optimization by Nature-Inspired Algorithms*. Springer, Singapore, 2018.

[5] Yang, Xin-She. *Nature-Inspired Metaheuristic Algorithms*. Luniver Press, University of Cambridge, UK, 2010.

[6] Mirjalili, Seyedali, and Jin Song Dong. "Introduction to nature-inspired algorithms." In *Nature-Inspired Optimizers*, pp. 1–5. Springer, Cham, 2020.

[7] Goldberg, David E. *Genetic Algorithms in Search, Optimization, and Machine Learning*. Addison-Wesley, Reading, Boston, MA, 1989.

[8] Deb, Kalyanmoy. "Multi-objective optimisation using evolutionary algorithms: An introduction." In *Multi-Objective Evolutionary Optimisation for Product Design and Manufacturing*, pp. 3–34. Springer, London, 2011.

[9] Van Laarhoven, Peter J.M., and Emile H.L. Aarts. "Simulated annealing." In *Simulated Annealing: Theory and Applications*, pp. 7–15. Springer, Dordrecht, 1987.

[10] Bertsimas, Dimitris, and John Tsitsiklis. "Simulated annealing." *Statistical Science* 8, no. 1 (1993): 10–15.

[11] Rutenbar, Rob A. "Simulated annealing algorithms: An overview." *IEEE Circuits and Devices Magazine* 5, no. 1 (1989): 19–26.

[12] Xi, Bowei, Zhen Liu, Mukund Raghavachari, Cathy H. Xia, and Li Zhang. "A smart hill-climbing algorithm for application server configuration." In *Proceedings of the 13th International Conference on World Wide Web*, pp. 287–296. ACM, New York, NY, 2004.

[13] Tsamardinos, Ioannis, Laura E. Brown, and Constantin F. Aliferis. "The max-min hill-climbing Bayesian network structure learning algorithm." *Machine Learning* 65, no. 1 (2006): 31–78.

[14] Gallego, Ramon A., Rubén Romero, and Alcir J. Monticelli. "Tabu search algorithm for network synthesis." *IEEE Transactions on Power Systems* 15, no. 2 (2000): 490–495.

[15] Jaeggi, Daniel M., Geoffrey T. Parks, Timoleon Kipouros, and P. John Clarkson. "The development of a multi-objective Tabu Search algorithm for continuous optimisation problems." *European Journal of Operational Research* 185, no. 3 (2008): 1192–1212.

[16] Díaz, Eugenia, Javier Tuya, Raquel Blanco, and José Javier Dolado. "A tabu search algorithm for structural software testing." *Computers & Operations Research* 35, no. 10 (2008): 3052–3072.

[17] Jiao, LiCheng, Lin Li, RongHua Shang, Fang Liu, and Rustam Stolkin. "A novel selection evolutionary strategy for constrained optimization." *Information Sciences* 239 (2013): 122–141.

[18] Siu, Sammy, Sheng-Sung Yang, Chien-Min Lee, and Chia-Lu Ho. "Improving the back-propagation algorithm using evolutionary strategy." *IEEE Transactions on Circuits and Systems II: Express Briefs* 54, no. 2 (2007): 171–175.

[19] Sinha, Nidul, R. Chakrabarti, and P.K. Chattopadhyay. "Evolutionary programming techniques for economic load dispatch." *IEEE Transactions on Evolutionary Computation* 7, no. 1 (2003): 83–94.

[20] Lee, Chang-Yong, and Xin Yao. "Evolutionary programming using mutations based on the Lévy probability distribution." *IEEE Transactions on Evolutionary Computation* 8, no. 1 (2004): 1–13.

[21] Babatunde, Oluleye H., Leisa Armstrong, Jinsong Leng, and Dean Diepeveen. "A genetic algorithm-based feature selection." *International Journal of Electronics Communication and Computer Engineering* 5, no. 4 (2014): 899–905.

[22] Hu, Xiao-Bing, and Ezequiel Di Paolo. "Binary-representation-based genetic algorithm for aircraft arrival sequencing and scheduling." *IEEE Transactions on Intelligent Transportation Systems* 9, no. 2 (2008): 301–310.

[23] Houck, Christopher R., Jeff Joines, and Michael G. Kay. "A genetic algorithm for function optimization: A Matlab implementation." *NCSU-IE Technical Report* 95, no. 09 (1995): 1–10.

[24] He, Yaohua, and Chi-Wai Hui. "A binary coding genetic algorithm for multi-purpose process scheduling: A case study." *Chemical Engineering Science* 65, no. 16 (2010): 4816–4828.

[25] Katoch, Sourabh, Sumit Singh Chauhan, and Vijay Kumar. "A review on genetic algorithm: Past, present, and future." *Multimedia Tools and Applications* (2020): 1–36.

[26] Wu, Huawei, Seyed Amin Bagherzadeh, Annunziata D'Orazio, Navid Habibollahi, Arash Karimipour, Marjan Goodarzi, and Quang-Vu Bach. "Present a new multi objective optimization statistical Pareto frontier method composed of artificial neural network and multi objective genetic algorithm to improve the pipe flow hydrodynamic and thermal properties such as pressure drop and heat transfer coefficient for non-Newtonian binary fluids." *Physica A: Statistical Mechanics and its Applications* 535 (2019): 122409, ISSN 0378-4371.

[27] Patle, B.K., D.R.K. Parhi, Anne Jagadeesh, and Sunil Kumar Kashyap. "Matrix-binary codes based genetic algorithm for path planning of mobile robot." *Computers & Electrical Engineering* 67 (2018): 708–728.

[28] Ali, Mostafa Z., Noor H. Awad, Ponnuthurai N. Suganthan, Ali M. Shatnawi, and Robert G. Reynolds. "An improved class of real-coded genetic algorithms for numerical optimization☆." *Neurocomputing* 275 (2018): 155–166.

[29] Mohamed, Ali Wagdy, and Ponnuthurai Nagaratnam Suganthan. "Real-parameter unconstrained optimization based on enhanced fitness-adaptive differential evolution algorithm with novel mutation." *Soft Computing* 22, no. 10 (2018): 3215–3235.

[30] Hellwig, Michael, and Hans-Georg Beyer. "A matrix adaptation evolution strategy for constrained real-parameter optimization." In *2018 IEEE Congress on Evolutionary Computation (CEC)*, pp. 1–8. IEEE, Manhattan, NY, 2018.

[31] Bora, Teodoro Cardoso, Viviana Cocco Mariani, and Leandro dos Santos Coelho. "Multi-objective optimization of the environmental-economic dispatch with reinforcement learning based on non-dominated sorting genetic algorithm." *Applied Thermal Engineering* 146 (2019): 688–700.

[32] Abdelsalam, Ali M., and M.A. El-Shorbagy. "Optimization of wind turbines siting in a wind farm using genetic algorithm based local search." *Renewable Energy* 123 (2018): 748–755.

[33] Mirjalili, Seyedali. "Genetic algorithm." In *Evolutionary Algorithms and Neural Networks*, pp. 43–55. Springer, Cham, 2019.

[34] Chatterjee, Sankhadeep, Rhitaban Nag, Nilanjan Dey, and Amira S. Ashour. "Efficient economic profit maximization: Genetic algorithm based approach." In *Smart Trends in Systems, Security and Sustainability*, pp. 307–318. Springer, Singapore, 2018.

[35] Mirjalili, Seyedali, Jin Song Dong, Ali Safa Sadiq, and Hossam Faris. "Genetic algorithm: Theory, literature review, and application in image reconstruction." *Nature-Inspired Optimizers* (2020): 69–85.

[36] Wang, Dongshu, Dapei Tan, and Lei Liu. "Particle swarm optimization algorithm: An overview." *Soft Computing* 22, no. 2 (2018): 387–408.

[37] Jain, N.K., Uma Nangia, and Jyoti Jain. "A review of particle swarm optimization." *Journal of the Institution of Engineers (India): Series B* 99, no. 4 (2018): 407–411.

[38] Bansal, Jagdish Chand. "Particle swarm optimization." In *Evolutionary and Swarm Intelligence Algorithms*, pp. 11–23. Springer, Cham, 2019.

[39] Piotrowski, Adam P., Jaroslaw J. Napiorkowski, and Agnieszka E. Piotrowska. "Population size in particle swarm optimization." *Swarm and Evolutionary Computation* 58 (2020): 100718.

[40] Jahandideh-Tehrani, Mahsa, Omid Bozorg-Haddad, and Hugo A. Loáiciga. "Application of particle swarm optimization to water management: An introduction and overview." *Environmental Monitoring and Assessment* 192, no. 5 (2020): 1–18.

[41] Kennedy, James, and Russell Eberhart. "Particle swarm optimization." In *Proceedings of ICNN'95-International Conference on Neural Networks*, vol. 4, pp. 1942–1948. IEEE, Manhattan, NY, 1995.

[42] Poli, Riccardo, James Kennedy, and Tim Blackwell. "Particle swarm optimization." *Swarm Intelligence* 1, no. 1 (2007): 33–57.

[43] Clerc, Maurice. *Particle Swarm Optimization*. Vol. 93. John Wiley & Sons, Hoboken, NJ, 2010.

[44] Shi, Yuhui, and Russell C. Eberhart. "Empirical study of particle swarm optimization." In *Proceedings of the 1999 Congress on Evolutionary Computation-CEC99 (Cat. No. 99TH8406)*, vol. 3, pp. 1945–1950. IEEE, Manhattan, NY, 1999.

[45] Liu, Junhong, and Jouni Lampinen. "A fuzzy adaptive differential evolution algorithm." *Soft Computing* 9, no. 6 (2005): 448–462.

[46] Qin, A. Kai, and Ponnuthurai N. Suganthan. "Self-adaptive differential evolution algorithm for numerical optimization." In *2005 IEEE Congress on Evolutionary Computation*, vol. 2, pp. 1785–1791. IEEE, Manhattan, NY, 2005.

[47] Civicioglu, Pinar, and Erkan Besdok. "Bernstain-search differential evolution algorithm for numerical function optimization." *Expert Systems with Applications* 138 (2019): 112831.

[48] Deng, Wu, Shifan Shang, Xing Cai, Huimin Zhao, Yingjie Song, and Junjie Xu. "An improved differential evolution algorithm and its application in optimization problem." *Soft Computing* 25, no. 7 (2021): 5277–5298.

[49] Deng, Wu, Hailong Liu, Junjie Xu, Huimin Zhao, and Yingjie Song. "An improved quantum-inspired differential evolution algorithm for deep belief network." *IEEE Transactions on Instrumentation and Measurement* 69, no. 10 (2020): 7319–7327.

[50] Deng, Wu, Hailong Liu, Junjie Xu, Huimin Zhao, and Yingjie Song. "An improved quantum-inspired differential evolution algorithm for deep belief network." *IEEE Transactions on Instrumentation and Measurement* 69, no. 10 (2020): 7319–7327.

[51] Emary, Eid, Hossam M. Zawbaa, Crina Grosan, and Abul Ella Hassenian. "Feature subset selection approach by gray-wolf optimization." In *Afro-European Conference for Industrial Advancement*, pp. 1–13. Springer, Cham, 2015.

[52] Emary, E., Waleed Yamany, Aboul Ella Hassanien, and Vaclav Snasel. "Multi-objective gray-wolf optimization for attribute reduction." *Procedia Computer Science* 65 (2015): 623–632.

[53] Singh, Narinder, and S.B. Singh. "A modified mean gray wolf optimization approach for benchmark and biomedical problems." *Evolutionary Bioinformatics* 13 (2017): 1176934317729413.

[54] Sun, Xiaodong, Changchang Hu, Gang Lei, Youguang Guo, and Jianguo Zhu. "State feedback control for a PM hub motor based on gray wolf optimization algorithm." *IEEE Transactions on Power Electronics* 35, no. 1 (2019): 1136–1146.

[55] Zareie, Ahmad, Amir Sheikhahmadi, and Mahdi Jalili. "Identification of influential users in social network using gray wolf optimization algorithm." *Expert Systems with Applications* 142 (2020): 112971.
[56] Goli, Alireza, Hassan Khademi Zare, Reza Tavakkoli Moghaddam, and Ahmad Sadeghieh. "An improved artificial intelligence based on gray wolf optimization and cultural algorithm to predict demand for dairy products: A case study." *IJIMAI* 5, no. 6 (2019): 15–22.
[57] Singh, Priyanka, Pragya Dwivedi, and Vibhor Kant. "A hybrid method based on neural network and improved environmental adaptation method using Controlled Gaussian Mutation with real parameter for short-term load forecasting." *Energy* 174 (2019): 460–477.
[58] Singh, Tribhuvan, Ranvijay Singh, and Krishn Kumar Mishra. "Software cost estimation using environmental adaptation method." *Procedia Computer Science* 143 (2018): 325–332.
[59] Singh, Navjot, K.K. Mishra, and Sanjiv Bhatia. "SEAM-an improved environmental adaptation method with real parameter coding for salient object detection." *Multimedia Tools and Applications* 79, no. 19 (2020): 12995–13010.
[60] Singh, Tribhuvan, Krishn Kumar Mishra, and Ranvijay. "A variant of EAM to uncover community structure in complex networks." *International Journal of Bio-Inspired Computation* 16, no. 2 (2020): 102–110.
[61] Tiwari, Shailesh, Anuj Kumar, Akash Punhani, and K.K. Mishra. "Modified environmental adaptation method and its application in test case generation." *Journal of Engineering Science & Technology Review* 12, no. 6 (2019).
[62] Singh, Tribhuvan, and Krishn Kumar Mishra. "Multiobjective environmental adaptation method for solving environmental/economic dispatch problem." *Evolutionary Intelligence* 12, no. 2 (2019): 305–319.
[63] Singh, Tribhuvan, Ranvijay Singh, and Krishn Kumar Mishra. "Software cost estimation using environmental adaptation method." *Procedia Computer Science* 143 (2018): 325–332.
[64] Sharma, Bhavna, Ravi Prakash, Shailesh Tiwari, and K.K. Mishra. "A variant of environmental adaptation method with real parameter encoding and its application in economic load dispatch problem." *Applied Intelligence* 47, no. 2 (2017): 409–429.
[65] Cotter, Neil E. "The Stone-Weierstrass theorem and its application to neural networks." *IEEE Transactions on Neural Networks* 1, no. 4 (1990): 290–295.
[66] Mishra, K.K. "New bio-inspired optimization algorithm." *Ph. D. Thesis, MNNIT Allahabad, India*, 2013.
[67] Dorigo, Marco, Vittorio Maniezzo, and Alberto Colorni. "Positive feedback as a search strategy." *Tech Rep.*, 91-016, Dip Elettronica, Politecnico di Milano, Italy, 1991.
[68] Dorigo, Marco, and Mauro Birattari. "Ant colony optimization." In *Encyclopedia of Machine Learning*, p. 39. Springer, Boston, 2010.
[69] Eberhart, Russell, and James Kennedy. "A new optimizer using particle swarm theory." In *MHS'95. Proceedings of the Sixth International Symposium on Micro Machine and Human Science*, pp. 39–43. IEEE, Manhattan, NY, 1995.
[70] Eberhart, Russell, and James Kennedy. "Particle swarm optimization." In *Proceedings of the IEEE International Conference on Neural Networks*, vol. 4, pp. 1942–1948, Perth, WA, Australia, 1995.
[71] Bonabeau, Eric, Directeur de Recherches Du Fnrs Marco, Marco Dorigo, Guy Théraulaz, and Guy Theraulaz. *Swarm Intelligence: From Natural to Artificial Systems*. No. 1. Oxford University Press, New York, NY, 1999.
[72] Dorigo, Marco, Mauro Birattari, and Thomas Stutzle. "Ant colony optimization." *IEEE Computational Intelligence Magazine* 1, no. 4 (2006): 28–39.
[73] Engelbrecht, Andries P. *Fundamentals of Computational Swarm Intelligence*. John Wiley & Sons, Inc., Hoboken, NJ, 2006.
[74] Blum, Christian, and Xiaodong Li. "Swarm intelligence in optimization." In *Swarm Intelligence*, pp. 43–85. Springer, Berlin, Heidelberg, 2008.

[75] Bonabeau, Eric, and Christopher Meyer. "Swarm intelligence." *Harvard Business Review* 79, no. 5 (2001): 106–114.

[76] Agrawal, Vinay, Vinay Chandwani, and Ravindra Nagar. "Swarm intelligence assisted optimization in structural engineering: A review." *INROADS-An International Journal of Jaipur National University* 3, no. 1s (2014): 173–180.

[77] Chetty, Sivashan, and Aderemi O. Adewumi. "Comparison study of swarm intelligence techniques for the annual crop planning problem." *IEEE Transactions on Evolutionary Computation* 18, no. 2 (2013): 258–268.

[78] Colorni, Alberto, Marco Dorigo, and Vittorio Maniezzo. "Distributed optimization by ant colonies." In *Proceedings of the First European Conference on Artificial Life*, vol. 142, pp. 134–142, Cambridge, MA, 1991.

[79] Dorigo, Marco, and Christian Blum. "Ant colony optimization theory: A survey." *Theoretical Computer Science* 344, no. 2–3 (2005): 243–278.

[80] Dorigo, Marco, Vittorio Maniezzo, and Alberto Colorni. "Ant system: Optimization by a colony of cooperating agents." *IEEE Transactions on Systems, Man, and Cybernetics, Part B (Cybernetics)* 26, no. 1 (1996): 29–41.

[81] Dorigo, M., and T. Stützle. *Ant Colony Optimization*, pp. 261–264. MIT Press, Cambridge, MA, 2004, 300 pp. (2005).

[82] Dorigo, Marco, and Luca Maria Gambardella. "Ant colony system: A cooperative learning approach to the traveling salesman problem." *IEEE Transactions on Evolutionary Computation* 1, no. 1 (1997): 53–66.

[83] Chira, Camelia, D. Dumitrescu, and Camelia-Mihaela Pintea. "Sensitive ant model for combinatorial optimization." In *Proceedings of the 12th WSEAS International Conference on Computers*, Stevens Point, WI, 2008.

[84] Gutjahr, Walter J. "A graph-based ant system and its convergence." *Future Generation Computer Systems* 16, no. 8 (2000): 873–888.

[85] Karaboga, Dervis. *An Idea based on Honey Bee Swarm for Numerical Optimization*. Vol. 200. Technical report-tr06, Erciyes university, engineering faculty, computer engineering department, 2005.

[86] Basturk, Bahriye. "An artificial bee colony (ABC) algorithm for numeric function optimization." In *IEEE Swarm Intelligence Symposium, Indianapolis, IN, USA*, 2006.

[87] Karaboga, Dervis, and Bahriye Basturk. "A powerful and efficient algorithm for numerical function optimization: Artificial bee colony (ABC) algorithm." *Journal of Global Optimization* 39, no. 3 (2007): 459–471.

[88] Schwefel, Hans-Paul. "On the evolution of evolutionary computation." *Computational Intelligence: Imitating Life* (1994): 116–124.

[89] Karaboga, D., and B. Basturk. "Advances in soft computing: Foundations of fuzzy logic and soft computing." *Artificial Bee Colony (ABC) Optimization Algorithm for Solving Constrained Optimization Problems, LNCS* 4529 (2007): 789–798.

[90] Basturk, Bahriye. "An artificial bee colony (ABC) algorithm for numeric function optimization." In *IEEE Swarm Intelligence Symposium, Indianapolis, IN, USA*, 2006.

[91] Karaboga, Dervis, and Bahriye Basturk. "On the performance of artificial bee colony (ABC) algorithm." *Applied Soft Computing* 8, no. 1 (2008): 687–697.

[92] Tereshko, Valery, and Andreas Loengarov. "Collective decision making in honey-bee foraging dynamics." *Computing and Information Systems* 9, no. 3 (2005): 1.

[93] Zou, Wenping, Yunlong Zhu, Hanning Chen, and Beiwei Zhang. "Solving multiobjective optimization problems using artificial bee colony algorithm." *Discrete Dynamics in Nature and Society* 2011 (2011).

[94] Gao, Wei-feng, and San-yang Liu. "A modified artificial bee colony algorithm." *Computers & Operations Research* 39, no. 3 (2012): 687–697.

[95] Lei, Xiujuan, Xu Huang, and Aidong Zhang. "Improved artificial bee colony algorithm and its application in data clustering." In *2010 IEEE Fifth International Conference*

on Bio-Inspired Computing: Theories and Applications (BIC-TA), pp. 514–521. IEEE, Piscataway, NJ, 2010.

[96] Narasimhan, Harikrishna. "Parallel artificial bee colony (PABC) algorithm." In *2009 World Congress on Nature & Biologically Inspired Computing (NaBIC)*, pp. 306–311. IEEE, Coimbatore, India, 2009.

[97] Banharnsakun, Anan, Tiranee Achalakul, and Booncharoen Sirinaovakul. "Artificial bee colony algorithm on distributed environments." In *2010 Second World Congress on Nature and Biologically Inspired Computing (NaBIC)*, pp. 13–18. IEEE, Kitakyushu, Japan, 2010.

[98] Karaboga, Dervis, and Bahriye Basturk. "Artificial bee colony (ABC) optimization algorithm for solving constrained optimization problems." In *International Fuzzy Systems Association World Congress*, pp. 789–798. Springer, Berlin, Heidelberg, 2007.

[99] Brajevic, Ivona, Milan Tuba, and Milos Subotic. "Improved artificial bee colony algorithm for constrained problems." In *Proceedings of the 11th WSEAS International Conference on Nural Networks and 11th WSEAS International Conference on Evolutionary Computing and 11th WSEAS International Conference on Fuzzy Systems*, pp. 185–190, World Scientific and Engineering Academy and Society (WSEAS), Stevens Point, WI, 2010.

[100] Mezura-Montes, Efrén, Mauricio Damián-Araoz, and Omar Cetina-Domíngez. "Smart flight and dynamic tolerances in the artificial bee colony for constrained optimization." In *IEEE Congress on Evolutionary Computation*, pp. 1–8. IEEE, Barcelona, Spain, 2010.

[101] Garg, N.K., Shimpi Singh Jadon, Harish Sharma, and D.K. Palwalia. "Gbest-artificial bee colony algorithm to solve load flow problem." In *Proceedings of the Third International Conference on Soft Computing for Problem Solving*, pp. 529–538. Springer, New Delhi, 2014.

[102] Pan, Quan-Ke, Ling Wang, Jun-Qing Li, and Jun-Hua Duan. "A novel discrete artificial bee colony algorithm for the hybrid flowshop scheduling problem with makespan minimisation." *Omega* 45 (2014): 42–56.

[103] Poli, Riccardo, James Kennedy, and Tim Blackwell. "Particle swarm optimization." *Swarm Intelligence* 1, no. 1 (2007): 33–57.

[104] Han, Jian-Wen, Hou-Feng Zheng, Yong Cui, Liang-Dan Sun, Dong-Qing Ye, Zhi Hu, Jin-Hua Xu et al. "Genome-wide association study in a Chinese Han population identifies nine new susceptibility loci for systemic lupus erythematosus." *Nature Genetics* 41, no. 11 (2009): 1234–1237.

[105] Arumugam, M. Senthil, M.V.C. Rao, and Aarthi Chandramohan. "A new and improved version of particle swarm optimization algorithm with global—local best parameters." *Knowledge and Information Systems* 16, no. 3 (2008): 331–357.

[106] Arumugam, M. Senthil, Machavaram Venkata Chalapathy Rao, and Alan W.C. Tan. "A novel and effective particle swarm optimization like algorithm with extrapolation technique." *Applied Soft Computing* 9, no. 1 (2009): 308–320.

[107] Shi, Yuhui, and Russell Eberhart. "*A modified particle swarm optimizer*." In 1998 IEEE international conference on evolutionary computation proceedings. IEEE world congress on computational intelligence (Cat. No. 98TH8360), pp. 69–73. IEEE, Piscataway, NJ, 1998.

[108] Shi, Yuhui. "Particle swarm optimization: Developments, applications and resources." In *Proceedings of the 2001 Congress on Evolutionary Computation (IEEE Cat. No. 01TH8546)*, vol. 1, pp. 81–86. IEEE, Piscataway, NJ, 2001.

[109] Clerc, Maurice, and James Kennedy. "The particle swarm-explosion, stability, and convergence in a multidimensional complex space." *IEEE Transactions on Evolutionary Computation* 6, no. 1 (2002): 58–73.

[110] Clerc, Maurice. "The swarm and the queen: Towards a deterministic and adaptive particle swarm optimization." In *Proceedings of the 1999 Congress on Evolutionary Computation-CEC99 (Cat. No. 99TH8406)*, vol. 3, pp. 1951–1957. IEEE, Piscataway, NJ, 1999.

[111] Shi, Yuhui, and Russell C. Eberhart. "Fuzzy adaptive particle swarm optimization." In *Proceedings of the 2001 Congress on Evolutionary Computation (IEEE Cat. No. 01TH8546)*, vol. 1, pp. 101–106. IEEE, Piscataway, NJ, 2001.

[112] Clerc, Maurice. "The swarm and the queen: Towards a deterministic and adaptive particle swarm optimization." In *Proceedings of the 1999 Congress on Evolutionary Computation-CEC99 (Cat. No. 99TH8406)*, vol. 3, pp. 1951–1957. IEEE, Piscataway, NJ, 1999.

[113] Gao, Fang, Gang Cui, Qiang Zhao, and Hongwei Liu. "Application of improved discrete particle swarm algorithm in partner selection of virtual enterprise." *International Journal of Computer Science and Network Security* 6, no. 3A (2006): 208–212.

[114] Premalatha, K., and A.M. Natarajan. "Discrete PSO with GA operators for document clustering." *International Journal of Recent Trends in Engineering* 1, no. 1 (2009): 20.

[115] Parsopoulos, Konstantinos E., and Michael N. Vrahatis. "Recent approaches to global optimization problems through particle swarm optimization." *Natural Computing* 1, no. 2 (2002): 235–306.

[116] Laskari, Elena C., Konstantinos E. Parsopoulos, and Michael N. Vrahatis. "Particle swarm optimization for integer programming." In *Proceedings of the 2002 Congress on Evolutionary Computation. CEC'02 (Cat. No. 02TH8600)*, vol. 2, pp. 1582–1587. IEEE, Piscataway, NJ, 2002.

[117] Shayeghi, H., M. Mahdavi, and A. Bagheri. "Discrete PSO algorithm based optimization of transmission lines loading in TNEP problem." *Energy Conversion and Management* 51, no. 1 (2010): 112–121.

[118] Yang, Xin-She, and Xingshi He. "Bat algorithm: Literature review and applications." *International Journal of Bio-Inspired Computation* 5, no. 3 (2013): 141–149.

[119] Yang, Xin-She. "Bat algorithm for multi-objective optimisation." *International Journal of Bio-Inspired Computation* 3, no. 5 (2011): 267–274.

[120] Mirjalili, Seyedali, Seyed Mohammad Mirjalili, and Xin-She Yang. "Binary bat algorithm." *Neural Computing and Applications* 25, no. 3 (2014): 663–681.

[121] Gandomi, Amir Hossein, Xin-She Yang, Amir Hossein Alavi, and Siamak Talatahari. "Bat algorithm for constrained optimization tasks." *Neural Computing and Applications* 22, no. 6 (2013): 1239–1255.

[122] Cui, Zhihua, Feixiang Li, and Wensheng Zhang. "Bat algorithm with principal component analysis." *International Journal of Machine Learning and Cybernetics* 10, no. 3 (2019): 603–622.

[123] Bajaj, Anu, and Om P. Sangwan. "Test case prioritization using bat algorithm." *Recent Advances in Computer Science and Communications (Formerly: Recent Patents on Computer Science)* 14, no. 2 (2021): 593–598.

[124] Chawla, Mridul, and Manoj Duhan. "Bat algorithm: A survey of the state-of-the-art." *Applied Artificial Intelligence* 29, no. 6 (2015): 617–634.

[125] Srivastava, Praveen Ranjan, Amit Bidwai, Anam Khan, Kritika Rathore, Rohit Sharma, and Xin She Yang. "An empirical study of test effort estimation based on bat algorithm." *International Journal of Bio-Inspired Computation* 6, no. 1 (2014): 57–70.

[126] Srivastava, Praveen Ranjan, and Tai-hoon Kim. "Application of genetic algorithm in software testing." *International Journal of Software Engineering and Its Applications* 3, no. 4 (2009): 87–96.

[127] Gou, Xiaodong, Tingting Huang, Shunkun Yang, Mengxuan Su, and Fuping Zeng. "Optimized differential evolution algorithm for software testing." *International Journal of Computational Intelligence Systems* 12, no. 1 (2018): 215–226.

[128] McMinn, Phil. "Search-based software testing: Past, present and future." In *2011 IEEE Fourth International Conference on Software Testing, Verification and Validation Workshops*, pp. 153–163. IEEE, Piscataway, NJ, 2011.

[129] Babamir, Faezeh Sadat, Alireza Hatamizadeh, Seyed Mehrdad Babamir, Mehdi Dabbaghian, and Ali Norouzi. "Application of genetic algorithm in automatic software

testing." In *International Conference on Networked Digital Technologies*, pp. 545–552. Springer, Berlin, Heidelberg, 2010.

[130] Bajaj, Anu, and Om Prakash Sangwan. "Tri-level regression testing using nature-inspired algorithms." *Innovations in Systems and Software Engineering* 17, no. 1 (2021): 1–16.

[131] Rosero, Raúl H., Omar S. Gómez, and Glen Rodríguez. "15 years of software regression testing techniques—a survey." *International Journal of Software Engineering and Knowledge Engineering* 26, no. 05 (2016): 675–689.

[132] Liang, Wei, Laibin Zhang, and Mingda Wang. "The chaos differential evolution optimization algorithm and its application to support vector regression machine." *Journal of Software* 6, no. 7 (2011): 1297–1304.

[133] Wang, Jianjun, Li Li, Dongxiao Niu, and Zhongfu Tan. "An annual load forecasting model based on support vector regression with differential evolution algorithm." *Applied Energy* 94 (2012): 65–70.

[134] Elsayed, Saber M., Ruhul A. Sarker, and Daryl L. Essam. "An improved self-adaptive differential evolution algorithm for optimization problems." *IEEE Transactions on Industrial Informatics* 9, no. 1 (2012): 89–99.

[135] Konsaard, Patipat, and Lachana Ramingwong. "Total coverage based regression test case prioritization using genetic algorithm." In *2015 12th International Conference on Electrical Engineering/Electronics, Computer, Telecommunications and Information Technology (ECTI-CON)*, pp. 1–6. IEEE, Piscataway, NJ, 2015.

[136] Mishra, Pradipta Kumar, and B.K.S.S. Pattanaik. "Analysis of test case prioritization in regression testing using genetic algorithm." *International Journal of Computer Applications* 75, no. 8 (2013).

[137] Saraswat, Birendra Kumar, Binayak Parashar, Arun Kumar Takuli, and Samender Singh. "Case study for regression testing using genetic algorithm." In *Journal of Physics: Conference Series* 2007, no. 1 (2021): 012060. IOP Publishing, 2021.

[138] Kaur, Arvinder, and Shubhra Goyal. "A genetic algorithm for fault based regression test case prioritization." *International Journal of Computer Applications* 32, no. 8 (2011): 975–8887.

[139] Joseph, A.K., and G. Radhamani. "Hybrid test case optimization approach using genetic algorithm with adaptive neuro fuzzy inference system for regression testing." *Journal of Testing and Evaluation* 45, no. 6 (2017): 2283–2293.

[140] Kaur, Arvinder, and Shubhra Goyal. "A genetic algorithm for regression test case prioritization using code coverage." *International Journal on Computer Science and Engineering* 3, no. 5 (2011): 1839–1847.

[141] Sharma, Chayanika, Sangeeta Sabharwal, and Ritu Sibal. "A survey on software testing techniques using genetic algorithm." *arXiv preprint arXiv:1411.1154* (2014).

[142] Windisch, Andreas, Stefan Wappler, and Joachim Wegener. "Applying particle swarm optimization to software testing." In *Proceedings of the 9th Annual Conference on Genetic and Evolutionary Computation*, pp. 1121–1128, Association for Computing Machinery, New York, NY, 2007.

[143] Chen, Xiang, Qing Gu, Jingxian Qi, and Daoxu Chen. "Applying particle swarm optimization to pairwise testing." In *2010 IEEE 34th Annual Computer Software and Applications Conference*, pp. 107–116. IEEE, Piscataway, NJ, 2010.

[144] Arora, Deepti, and Anurag Singh Baghel. "Application of genetic algorithm and particle swarm optimization in software testing." *IOSR Journal of Computer Engineering* 17, no. 1 (2015): 75–78.

[145] Kaur, Arvinder, and Divya Bhatt. "Hybrid particle swarm optimization for regression testing." *International Journal on Computer Science and Engineering* 3, no. 5 (2011): 1815–1824.

[146] Hla, Khin Haymar Saw, YoungSik Choi, and Jong Sou Park. "Applying particle swarm optimization to prioritizing test cases for embedded real time software retesting." In *2008 IEEE 8th International Conference on Computer and Information Technology Workshops*, pp. 527–532. IEEE, Piscataway, NJ, 2008.

[147] Meissner, Michael, Michael Schmuker, and Gisbert Schneider. "Optimized particle swarm optimization (OPSO) and its application to artificial neural network training." *BMC Bioinformatics* 7, no. 1 (2006): 1–11.

[148] Tiwari, Shailesh, K.K. Mishra, and Arun Kumar Misra. "Test case generation for modified code using a variant of particle swarm optimization (PSO) algorithm." In *2013 10th International Conference on Information Technology: New Generations: New generations*, pp. 363–368. IEEE, Piscataway, NJ, 2013.

[149] Zhang, Yudong, Shuihua Wang, and Genlin Ji. "A comprehensive survey on particle swarm optimization algorithm and its applications." *Mathematical Problems in Engineering* 2015 (2015).

[150] Yuan, Yingchun, Xiaoping Li, Qian Wang, and Xia Zhu. "Deadline division-based heuristic for cost optimization in workflow scheduling." *Information Sciences* 179, no. 15 (2009): 2562–2575.

[151] Wu, Zhangjun, Zhiwei Ni, Lichuan Gu, and Xiao Liu. "A revised discrete particle swarm optimization for cloud workflow scheduling." In *2010 International Conference on Computational Intelligence and Security*, pp. 184–188. IEEE, Piscataway, NJ, 2010.

[152] Chen, Wei-Neng, and Jun Zhang. "An ant colony optimization approach to a grid workflow scheduling problem with various QoS requirements." *IEEE Transactions on Systems, Man, and Cybernetics, Part C (Applications and Reviews)* 39, no. 1 (2008): 29–43.

[153] Bilgaiyan, Saurabh, Santwana Sagnika, and Madhabananda Das. "Workflow scheduling in cloud computing environment using cat swarm optimization." In *2014 IEEE International Advance Computing Conference (IACC)*, pp. 680–685. IEEE, Piscataway, NJ, 2014.

[154] Singh, Lovejit, and Sarbjeet Singh. "A survey of workflow scheduling algorithms and research issues." *International Journal of Computer Applications* 74, no. 15 (2013).

[155] Xu, Rongbin, Yeguo Wang, Yongliang Cheng, Yuanwei Zhu, Ying Xie, Abubakar Sadiq Sani, and Dong Yuan. "Improved particle swarm optimization based workflow scheduling in cloud-fog environment." In *International Conference on Business Process Management*, pp. 337–347. Springer, Cham, 2018.

[156] Wu, Fuhui, Qingbo Wu, and Yusong Tan. "Workflow scheduling in cloud: A survey." *The Journal of Supercomputing* 71, no. 9 (2015): 3373–3418.

[157] Hosseinzadeh, Mehdi, Marwan Yassin Ghafour, Hawkar Kamaran Hama, Bay Vo, and Afsane Khoshnevis. "Multi-objective task and workflow scheduling approaches in cloud computing: A comprehensive review." *Journal of Grid Computing* (2020): 1–30.

[158] Alkhanak, Ehab Nabiel, Sai Peck Lee, Reza Rezaei, and Reza Meimandi Parizi. "Cost optimization approaches for scientific workflow scheduling in cloud and grid computing: A review, classifications, and open issues." *Journal of Systems and Software* 113 (2016): 1–26.

[159] Arya, Lokesh Kumar, and Amandeep Verma. "Workflow scheduling algorithms in cloud environment-A survey." In *2014 Recent Advances in Engineering and Computational Sciences (RAECS)*, pp. 1–4. IEEE, Piscataway, NJ, 2014.

[160] Farid, Mazen, Rohaya Latip, Masnida Hussin, and Nor Asilah Wati Abdul Hamid. "A survey on QoS requirements based on particle swarm optimization scheduling techniques for workflow scheduling in cloud computing." *Symmetry* 12, no. 4 (2020): 551.

[161] Ram, Satya Deo K., Shashank Srivastava, and Krishn Kumar Mishra. "A variant of teaching-learning-based optimization and its application for minimizing the cost of workflow execution in the cloud computing." *Concurrency and Computation: Practice and Experience* (2021): e6425.

[162] Wu, Hao, Xin Chen, Xiaoyu Song, Chi Zhang, and He Guo. "Scheduling large-scale scientific workflow on virtual machines with different numbers of vCPUs." *The Journal of Supercomputing* 77, no. 1 (2021): 679–710.

[163] Abdel-Basset, Mohamed, Reda Mohamed, Ripon K. Chakrabortty, Karam Sallam, and Michael J. Ryan. "An efficient teaching-learning-based optimization algorithm for parameters identification of photovoltaic models: Analysis and validations." *Energy Conversion and Management* 227 (2021): 113614.

[164] Mareli, M., and B. Twala. "An adaptive Cuckoo search algorithm for optimisation." *Applied Computing and Informatics* 14, no. 2 (2018): 107–115.
[165] Yang, Xin-She, and Suash Deb. "Cuckoo search: Recent advances and applications." *Neural Computing and Applications* 24, no. 1 (2014): 169–174.
[166] Gandomi, Amir Hossein, Xin-She Yang, and Amir Hossein Alavi. "Cuckoo search algorithm: A metaheuristic approach to solve structural optimization problems." *Engineering with Computers* 29, no. 1 (2013): 17–35.
[167] Valian, Ehsan, Shahram Mohanna, and Saeed Tavakoli. "Improved cuckoo search algorithm for global optimization." *International Journal of Communications and Information Technology* 1, no. 1 (2011): 31–44.
[168] Mahdavi, Mehrdad, Mohammad Fesanghary, and Ebrahim Damangir. "An improved harmony search algorithm for solving optimization problems." *Applied Mathematics and Computation* 188, no. 2 (2007): 1567–1579.
[169] Lee, Kang Seok, and Zong Woo Geem. "A new structural optimization method based on the harmony search algorithm." *Computers & Structures* 82, no. 9–10 (2004): 781–798.
[170] Geem, Zong Woo, ed. *Music-Inspired Harmony Search Algorithm: Theory and Applications*. Vol. 191. Springer, Berlin, Heidelberg, 2009.
[171] Manjarres, Diana, Itziar Landa-Torres, Sergio Gil-Lopez, Javier Del Ser, Miren Nekane Bilbao, Sancho Salcedo-Sanz, and Zong Woo Geem. "A survey on applications of the harmony search algorithm." *Engineering Applications of Artificial Intelligence* 26, no. 8 (2013): 1818–1831.
[172] Geem, Zong Woo. "Recent advances in harmony search algorithm." *Studies in Computational Intelligence* 270, ISBN 978-3-642-04316-1, e-ISBN 978-3-642-04317-8, DOI: 10.1007/978-3-642-04317-8, ISSN 1860-949X, Springer-Verlag Berlin Heidelberg, (2010).
[173] Yang, Xin-She. "Harmony search as a metaheuristic algorithm." In *Music-Inspired Harmony Search Algorithm*, pp. 1–14. Springer, Berlin, Heidelberg, 2009.
[174] Yang, Xin-She, and Xingshi He. "Firefly algorithm: Recent advances and applications." *International Journal of Swarm Intelligence* 1, no. 1 (2013): 36–50.
[175] Yang, Xin-She. "Firefly algorithm, stochastic test functions and design optimisation." *International Journal of Bio-Inspired Computation* 2, no. 2 (2010): 78–84.
[176] Gandomi, Amir H., X-S. Yang, Siamak Talatahari, and Amir Hossein Alavi. "Firefly algorithm with chaos." *Communications in Nonlinear Science and Numerical Simulation* 18, no. 1 (2013): 89–98.
[177] Yang, Xin-She. "Firefly algorithm, Levy flights and global optimization." In *Research and Development in Intelligent Systems XXVI*, pp. 209–218. Springer, London, 2010.
[178] Farahani, S.M., Azam A. Abshouri, Babak Nasiri, and M.R. Meybodi. "A Gaussian firefly algorithm." *International Journal of Machine Learning and Computing* 1, no. 5 (2011): 448.
[179] Erol, Osman K., and Ibrahim Eksin. "A new optimization method: Big bang—big crunch." *Advances in Engineering Software* 37, no. 2 (2006): 106–111.
[180] Camp, Charles V. "Design of space trusses using big bang—big crunch optimization." *Journal of Structural Engineering* 133, no. 7 (2007): 999–1008.
[181] Kaveh, Ali, and S. Talatahari. "Size optimization of space trusses using big bang—big crunch algorithm." *Computers & Structures* 87, no. 17–18 (2009): 1129–1140.
[182] Alatas, Bilal. "Uniform big bang—chaotic big crunch optimization." *Communications in Nonlinear Science and Numerical Simulation* 16, no. 9 (2011): 3696–3703.
[183] Prayogo, Doddy, Min-Yuan Cheng, Yu-Wei Wu, Albertus Arief Herdany, and Handy Prayogo. "Differential big bang-big crunch algorithm for construction-engineering design optimization." *Automation in Construction* 85 (2018): 290–304.
[184] Mirjalili, Seyedali, and Andrew Lewis. "The whale optimization algorithm." *Advances in Engineering Software* 95 (2016): 51–67.
[185] Gharehchopogh, Farhad Soleimanian, and Hojjat Gholizadeh. "A comprehensive survey: Whale optimization algorithm and its applications." *Swarm and Evolutionary Computation* 48 (2019): 1–24.

[186] Aljarah, Ibrahim, Hossam Faris, and Seyedali Mirjalili. "Optimizing connection weights in neural networks using the whale optimization algorithm." *Soft Computing* 22, no. 1 (2018): 1–15.
[187] Kaur, Gaganpreet, and Sankalap Arora. "Chaotic whale optimization algorithm." *Journal of Computational Design and Engineering* 5, no. 3 (2018): 275–284.
[188] Mafarja, Majdi M., and Seyedali Mirjalili. "Hybrid whale optimization algorithm with simulated annealing for feature selection." *Neurocomputing* 260 (2017): 302–312.
[189] Heidari, Ali Asghar, Seyedali Mirjalili, Hossam Faris, Ibrahim Aljarah, Majdi Mafarja, and Huiling Chen. "Harris hawks optimization: Algorithm and applications." *Future Generation Computer Systems* 97 (2019): 849–872.
[190] Too, Jingwei, Abdul Rahim Abdullah, and Norhashimah Mohd Saad. "A new quadratic binary Harris hawk optimization for feature selection." *Electronics* 8, no. 10 (2019): 1130.
[191] Birogul, Serdar. "Hybrid Harris hawk optimization based on differential evolution (HHODE) algorithm for optimal power flow problem." *IEEE Access* 7 (2019): 184468–184488.
[192] Mahapatra, Sheila, Bishwajit Dey, and Saurav Raj. "A novel ameliorated Harris hawk optimizer for solving complex engineering optimization problems." *International Journal of Intelligent Systems* 36, no. 12 (2021): 7641–7681.
[193] Zheng-Ming, G.A.O., Z.H.A.O. Juan, H.U. Yu-Rong, and Hua-Feng Chen. "The improved Harris hawk optimization algorithm with the Tent map." In *2019 3rd International Conference on Electronic Information Technology and Computer Engineering (EITCE)*, pp. 336–339. IEEE, Piscataway, NJ, 2019.
[194] Hatamlou, Abdolreza. "Black hole: A new heuristic optimization approach for data clustering." *Information Sciences* 222 (2013): 175–184.
[195] Piotrowski, Adam P., Jaroslaw J. Napiorkowski, and Pawel M. Rowinski. "How novel is the 'novel' black hole optimization approach?" *Information Sciences* 267 (2014): 191–200.
[196] Zhang, Junqi, Kun Liu, Ying Tan, and Xingui He. "Random black hole particle swarm optimization and its application." In *2008 International Conference on Neural Networks and Signal Processing*, pp. 359–365. IEEE, Piscataway, NJ, 2008.
[197] Bouchekara, H.R.E.H. "Optimal power flow using black-hole-based optimization approach." *Applied Soft Computing* 24 (2014): 879–888.

Questions

CHAPTER 1

MULTIPLE-CHOICE QUESTIONS

1. An example of a combinatorial optimization problem is
   a. the traveling salesperson problem.
   b. the curve-fitting problem.
   c. the binary search problem.
   d. None of the above

2. Domain D(X) of a discrete optimization problem contains
   a. an infinite solution.
   b. a finite number of solutions.
   c. Both a and b
   d. None of the above

3. A function evaluation is counted when
   a. the solution is updated.
   b. the iteration is changed.
   c. the fitness value of a solution is evaluated.
   d. None of the above

4. A benchmark function will be scalable if
   a. it is defined for a fixed dimension.
   b. it is adjustable with a dimension.
   c. Both a and b
   d. None of the above

5. The number of fitness evaluations and iterations are
   a. the same.
   b. possibly different.
   c. Do not know
   d. None of the above

6. Which of the following benchmark functions are used to check the performance of algorithms?
   a. CEC benchmarks
   b. BBOB benchmarks
   c. Both a and b
   d. None of the above

7. Which of the following is a type of encoding?
   a. Binary encoding
   b. Real-parameter encoding
   c. Both a and b
   d. None of the above

8. Which of the following optimization problem has multiple local and global optimal solutions?
   a. Unimodal problem
   b. Multimodal problem
   c. Both a and b
   d. None of the above

9. Which of the following problem contains only one global optimal solution?
   a. Unimodal problem
   b. Multimodal problem
   c. Both a and b
   d. None of the above

10. Which of the following algorithm is not based on real-parameter encoding?
    a. Binary genetic algorithm
    b. Binary grey wolf optimization
    c. Both a and b
    d. None of the above

SHORT-ANSWER QUESTIONS

1. Explain the binary encoding technique. Name some algorithms which use this technique.
2. Explain the difference between deterministic and randomized algorithms.
3. How will you check the performance of a nature-inspired algorithm?
4. For the given problem, in how many solutions, the search space should be divided to represent each solution by 16 bits?

   Minimize F(x) = Sin x, where 0 <=x <=3

5. In Question 4, what will be the actual value of 1100110011000001?
6. What are combinatorial optimization problems? Name some of them.
7. What are shifted benchmark functions?
8. What are scalable and fixed benchmark functions?
9. Name some frameworks used for checking the performance of new optimization algorithms.
10. How will you check whether a new algorithm is fast from existing algorithms or not?

LONG-ANSWER QUESTIONS

1. How can you create a new nature-inspired algorithm? How will you verify that its performance is better than other algorithms?
2. Explain the need to create randomized algorithms for solving optimization problems.
3. Define optimization problems. How can we classify these problems?
4. What is no-free-lunch theorem? Why is it useful?
5. Explain the difference between binary encoding and real-parameter encoding with a suitable example.

CHAPTER 2

MULTIPLE-CHOICE QUESTIONS

1. A genetic algorithm was developed by mapping
   a. the theory of Swarm Intelligence.
   b. the theory of natural selection and evolution.
   c. the theory of physics.
   d. None of the above

2. The purpose of the selection operator in a GA is
   a. to create diversity.
   b. to promote a good solution.
   c. to generate new solutions.
   d. to assign fitness.

3. Which of the following selection operator is used to solve multimodal problems?
   a. Tournament selection
   b. Roulette wheel selection
   c. Sharing fitness method
   d. None of the above

4. Which of the following theory establishes the theoretical foundations of GA?
   a. Theory of natural selection and evolution
   b. Schemata theory
   c. Both a and b
   d. None of these

5. What is the purpose of crossover and mutation operators?
   a. To create new solutions
   b. To select good solutions
   c. Both a and b
   d. None of the above

6. What is the purpose of the crossover operator?
   a. To perform exploitation
   b. To perform exploration
   c. Both a and b
   d. None of the above

7. What is the terminating condition in a GA?
   a. The number of iterations reaches the max value.
   b. The desired solution is obtained.
   c. Both a and b
   d. None of the above

8. What is an initial population in a GA?
   a. The number of iterations
   b. The number of generations
   c. Both a and b
   d. The number of solutions used to start the search

9. What is the other name of the solution in a GA?
   a. Iteration
   b. Chromosome
   c. Allie
   d. Generation

10. Which of the following functions will create 10 solutions having 4 dimensions?
    a. Ones(10,4)
    b. rand(10,4)
    c. Zeros(10,4)
    d. All of the above

11. Which of the following selection operators selects solutions according to their rank?
    a. Roulette wheel selection
    b. Tournament selection
    c. Rank-based selection
    d. Both a and c

12. In sharing fitness-based selection, sharing fitness of any solution decreases with an increase in
    a. sharing distance between solutions.
    b. the number of solutions close to the given solution.
    c. has no effect on a and b.
    d. both a and b.

13. The selection of a solution is made according to
    a. the position of solutions.
    b. the fitness of the solutions.
    c. both a and b.
    d. neither a nor b.

14. Which of the following is not an example of a selection operator in a GA?
    a. Tournament
    b. Rank-based
    c. One point
    d. All of the above

15. In a one-point crossover operator, the children become more diverse if a cut point is near to most significant bit of parents.
    a. True
    b. False
    c. Not related to the question
    d. Not connected

16. In a mutation operator, the diversity between parent and mutated child is dependent on the position of flip.
    a. True
    b. False
    c. Not related to the question
    d. Not connected

17. Which of the following is not a type of crossover in a GA?
    a. One point
    b. Uniform
    c. Rank
    d. Both a and b

18. The dimension of a solution represents
    a. the number of variables in the solution.
    b. the count of solutions.
    c. the decrease with the increase in the population.
    d. None of the above

19. What will happen if a selection is not used in a GA?
    a. It will not affect the normal working of a GA.
    b. It will not be able to identify the optimal solution.
    c. The direction of search cannot be predicted.
    d. Both b and c

20. A search space is divided into 2^{40} solutions. What will be the size of a chromosome?
    a. 2
    b. 4
    c. 20
    d. 40

SHORT-ANSWER QUESTIONS

1. Explain why the initial population in a GA is generated randomly?
2. Do encoding techniques affect the performance of a GA? Explain.
3. What are generation, iteration, and chromosome in a BGA?
4. Define *genotype* and *phenotype* with an example.
5. Write MATLAB commands to convert a binary number into real values in MATLAB.
6. What will happen if we forget to implement a selection operator in GA? Will it converge or not?
7. What is the difference between Roulette Wheel selection and rank-based selection in a GA?
8. Write MATLAB commands to create 20 solutions, each containing 20 bits. Explain the instructions.
9. Explain the uniform crossover operator with an example.
10. Explain the sharing-fitness method used in a GA.
11. Write down the name of the encoding techniques used in a GA.
12. What is the role of probability in the crossover and mutation operators of a GA?
13. Explain the GA with the help of a flow chart.
14. Write own the name of the crossover operators used in a GA.
15. Write down the name of mutation operators used in a BGA.
16. Write MATLAB command to initialize ten random solutions within the interval [−3,4].
17. Write the meaning of the following MATLAB commands:
    a. Ones(1,1)
    b. Zeros(10,20)
    c. 4*rand(10,10)-2;
    d. X=4*rand(20,20)-ones(20,20)
18. What is an elitist GA?
19. Which selection operator will you use to handle unimodal problems, and why?
20. A person wants to calculate the difference between fitness and sharing fitness in a GA. Explain whether they can obtain a negative value. Why?

LONG-ANSWER QUESTIONS

1. Explain the roulette wheel selection method in a GA with an example. What will happen if we do not normalize the total fitness between 0 to 1?
2. Explain the sharing-fitness method. Explain with an example.

3. Can you suggest any other method for crossover and mutation (not mentioned in this book)? If yes, how will you verify whether they are good or not?
4. Write a MATLAB function to accept a binary population of 10 solutions. Each solution contains 30 binary bits. Calculate the fitness of each solution. The detail of problem follows:

   Maximize $F(x_1, x_2, x_3) = x_1^2 + x_2^2 + x_3^2$,

   where $1 <= x_1 <= 5$, $2 <= x_2 <= 6$, $1 <= x_3 <= 7$.

   Assume that the equal number of bits represents each variable.
5. Assume that we do not apply a crossover operator in a GA. How will it affect the performance of GA? Would you please suggest some other method to replace the crossover?
6. What is the purpose of a crossover/cut point in the crossover operator? Can you suggest any other method for generating children using a crossover?
7. Do you feel that we have wholly implemented the theory of natural selection and evolution?

   Is there any chance to define any other operator which looks more natural?
8. Can we change the order of operators, that is, selection, crossover, and mutation? If yes, what changes do we have to make in the algorithm? If not, why not.
9. Explain the working of the BGA with an example.
10. What will happen when a GA starts converging on an optimal solution? How will you check this condition by checking the results only? What changes will reflect in different generations?
11. Write down the characteristics of a problem required for using a genetic algorithm. Using an example, show why it is essential to have a mutation operator in a genetic algorithm.
12. Describe how one-point crossover works in a GA. How it is different from a uniform crossover? Mention the condition when one-point crossover performs exploitation of search space.
13. Are GAs useful if we don't have a complete understanding of our objective function? Explain.
14. Consider a binary string (schema) 1*0**********, where * represents 0 or 1. Specify all conditions that must hold so that the instance of the schema will be destroyed by bit-flip mutation.

CHAPTER 3

MULTIPLE-CHOICE QUESTIONS

1. Is it possible to generate all solutions to a continuous optimization problem using binary encoding?
   a. Yes
   b. No
   c. Maybe
   d. None of the above

2. A roulette wheel selection operator can also be used in a real-parameter GA.
   a. Yes
   b. No
   c. Maybe
   d. None of the above

3. rand(3,3) will generate a matrix
   a. of size 3 × 3 having random numbers between 0 to 1.
   b. of size 3 × 3 having random numbers between 1 to 2.
   c. of size 3 × 3 having random numbers between 0 to 2.
   d. of size 3 × 3 having random numbers between 2 to 0.

4. Which problem or problems may occur when binary encoding is used for solving continuous optimization problems?
   a. Hamming cliff
   b. More computation
   c. Less coverage of search space
   d. All of the above

5. Is real-parameter encoding efficient in handling all optimization problems?
   a. Yes
   b. No
   c. Maybe
   d. None of the above

SHORT-ANSWER QUESTIONS

1. If ns = 5, then what will be the value of prob vector.
   prob = flipud([1:ns]'/sum([1:ns]))
2. Explain the arithmetic crossover operator.
3. Following is the code written by a programmer for implementing mutation in a population of 10 solutions having 3 variables in each solution. While writing the code, he did a mistake in this code. Can you please identify the mistake and explain whether it will affect the program for implementing mutation?
   % Mutate the population

```
nmut=6
mrow=sort(ceil(rand(1,nmut)*(Dim-1))+1);
mcol=ceil(rand(1,nmut)*popsize);
for ii=1:nmut
par(mrow(ii),mcol(ii))=(varhi-varlo)*rand+varlo;
% mutation
end% ii
```

4. Explain the working of a GA with the help of a flow chart.
5. Discuss why real-parameter encoding is required in continuous optimization problems.
6. What is a simulated binary crossover operator? What are the advantages of using this operator?

CHAPTER 4

MULTIPLE-CHOICE QUESTIONS

1. Which of the following is an evolutionary algorithm?
   a. Particle swarm optimization
   b. DE algorithm
   c. Grey wolf optimization algorithm
   d. None of the above

2. The order of operators in the DE algorithm is
   a. selection, crossover, and mutation.
   b. crossover, mutation, and selection.
   c. mutation, crossover, and selection.
   d. None of the above

3. Which of the following parameters is the control parameter in the DE algorithm?
   a. W
   b. C1
   c. C2
   d. F

4. For the selection operator of DE, select offspring solutions
   a. from the mutated vector and the target vector.
   b. from the target vector and the trail vector.
   c. from the trail vector and the mutated vector.
   d. None of the above

5. In DE, target vectors are mutated to generate a
   a. modified vector.
   b. donor vector.
   c. trail vector.
   d. None of the above

6. The convergence rate of a DE algorithm can be improved by
   a. defining new mutation strategies.
   b. updating crossover operator.
   c. parameter tuning.
   d. All of the above

7. A DE algorithm is mainly used for solving
   a. numerical optimization problems.
   b. discrete optimization problems.
   c. graph-related problems.
   d. None of the above

8. Which of the following solutions cannot be produced after selection?
   a. Donor vector
   b. Target vector
   c. Trail vector
   d. None of the above

9. Which of the following is not a type of crossover operator in DE?
   a. Binomial crossover
   b. Uniform crossover
   c. One-point crossover
   d. None of the above

10. The mutation operator of DE is applied on
    a. every parent solution.
    b. every offspring solution.
    c. selected solutions.
    d. None of the above

SHORT-ANSWER QUESTIONS

1. Differentiate between a GA and a DE algorithm.
2. Explain the purpose of the mutation operator.
3. What will be the effect of the selection operator in DE?
4. Write a MATLAB command to generate ten solutions of three dimensions in the interval [0,1].
5. Explain the purpose of the crossover operator in DE.
6. What is parameter tuning? How does it affect the performance of a DE algorithm?
7. What is the role of clamping in DE?
8. What will happen if we change mutation strategy of the DE?
9. What will happen if no selection operator is used in DE?
10. What will happen if the initial population is not initialized randomly?

LONG-ANSWER QUESTIONS

1. Why is mutation applied before crossover in DE, unlike GA? What are the shortcomings of GA that are handled by DE more effectively?
2. Apart from Equation 4.1, can you design another mutation strategy? If yes, show the example.

3. Describe the control parameters of DE in detail.
4. What are the applications of DE and GAs? In which scenarios will DE work better than a GA?
5. Will DE work effectively in dealing with binary optimization problems also?

CHAPTER 5

MULTIPLE-CHOICE QUESTIONS

1. Which of the following phenomena is designed to develop a PSO algorithm?
   a. Natural selection and evolution
   b. Birds flocking
   c. Adaptation
   d. None of the above

2. What is the role of gbest in a PSO?
   a. For implementing exploration in a PSO
   b. For implementing exploitation in a PSO
   c. Both a and b
   d. None of the above

3. What is the role of pbest in a PSO?
   a. For implementing exploration in a PSO
   b. For implementing exploitation in a PSO
   c. Both a and b
   d. None of the above

4. Which of the following techniques is used to make a particle feasible if it goes out of bound?
   a. Initialization
   b. Exploration
   c. Clamping
   d. None of the above

5. The pbest of each particle changes after
   a. each iteration.
   b. the fitness evaluation.
   c. the fitness of the new position is better than the previous position.
   d. None of the above

SHORT-ANSWER QUESTIONS

1. Explain the motivation behind introducing PSO?
2. What are pbest and gbest in PSO?

3. What are the exploration and exploitation properties?
4. Define self-adaptive algorithms?
5. What are the different parameters that control the performance of PSO?

LONG-ANSWER QUESTIONS

1. Explain PSO with its flow chart.
2. Why is PSO better than other optimization techniques?
3. State some real-time applications of PSO.

CHAPTER 6

MULTIPLE-CHOICE QUESTIONS

1. The GWO algorithm implements the theory of
   a. birds flocking.
   b. fish schooling.
   c. hunting process used by grey wolves.
   d. searching for a food source by a group of wolves.

2. The GWO algorithm is a(n)
   a. evolutionary algorithm.
   b. swarm intelligence algorithm.
   c. Both a and b
   d. None of the above

3. The position of the alpha wolf is identified by
   a. checking the fitness of each wolf in the population.
   b. checking the position of each wolf in the population.
   c. checking the velocity of each wolf in the population.
   d. None of the above

4. What is the purpose of omega wolf in the group?
   a. To guide alpha wolf
   b. To guide beta wolf
   c. To work according to the instructions received from alpha, beta, and delta wolves
   d. None of the above

5. Do we use a selection operator in GWO?
   a. Yes
   b. No
   c. Do not know
   d. None of the above

SHORT-ANSWER QUESTIONS

1. Explain whether GWO is an evolutionary algorithm or a swarm intelligence-based algorithm.
2. Explain the process of updating positions of the alpha, beta, and delta wolves.
3. State the differences between PSO and GWO.
4. How can we identify the positions of alpha, beta, and delta wolves in GWO?
5. What are the different parameters involved in updating the positions of the wolves in GWO?

LONG-ANSWER QUESTIONS

1. Illustrate the flow chart and pseudocode of GWO.
2. Show how GWO works using an example.
3. Compare the differential evolution, PSO, and GWO optimization algorithms.

CHAPTER 7

SHORT-ANSWER QUESTIONS:

1. What was the motivation behind creating the EAM?
2. What is the name of natural phenomenon that is used for creating the EAM?
3. Explain how IEAM is different from EAM.
4. What is the difference between binary IEAM and real-parameter IEAM?

CHAPTER 8

MULTIPLE-CHOICE QUESTIONS

1. Which of the following is point-to-point-based algorithm?
   a. Particle swarm optimization
   b. GA
   c. SA
   d. None of the above

2. Which of the following use *levy.s* flight concept for designing algorithm?
   a. ANO
   b. Ant–bee colony
   c. Cuckoo search algorithm
   d. None of the above

3. Which of the following is the latest version of ES?
   a. AES algorithms
   b. ES multi-algorithms
   c. CMA-ES algorithms
   d. None of the above

4. Which of the following is a population-based algorithm?
   a. Ant–bee colony
   b. Tabu search
   c. SA
   d. None of the above

5. Which of the following is a swarm intelligence–based algorithm?
   a. ES
   b. EP
   c. GA
   d. Bat algorithm

SHORT-ANSWER QUESTIONS

1. Why do we need nature-inspired algorithms?
2. Illustrate the taxonomy of nature-inspired algorithms?
3. What is point-to-point-based algorithm? Explain with an example.
4. What is the motivation behind evolutionary algorithms?
5. How is TLBO effective in improving the learner's knowledge?

LONG-ANSWER QUESTIONS

1. List the advantages of swarm intelligence–based algorithms.
2. Elaborate the optimization algorithm based on the evolution of the universe.
3. List the nature-inspired algorithms and their derivation from the natural phenomena.

CHAPTER 9

MULTIPLE-CHOICE QUESTIONS

1. Software testing is used to
   a. check errors in the software.
   b. improve the quality of software.
   c. Both a and b
   d. None of the above

2. Which of the following technique require to constrain solver for generating test cases?
   a. Random testing
   b. Dynamic testing
   c. Symbolic testing
   d. None of the above

3. Which of the following coverage criteria can be used to generate test cases in white-box testing?
   a. Statement coverage
   b. Branch coverage
   c. Path coverage
   d. All of the above

4. Which of the following tools can be used to check statement/block coverage in Linux?
   a. GCOV
   b. TRUCOV
   c. Both a and b
   d. None of the above

5. Which of the following test cases will not improve the quality of the software?
   a. A test case that produces the expected output
   b. A test case whose desired outcome is different from the actual one
   c. Both a and b
   d. None of the above

6. Which of the following testing techniques does not require the calculation of statement coverage?
   a. White-box testing
   b. Black-box testing
   c. Both a and b
   d. None of the above

7. If the testing budget is much less, which of the following techniques will be used to create test cases?
   a. Test cases will be generated to meet the requirement of the software.
   b. Test cases will be generated by the metaheuristic technique.
   c. Both a and b
   d. None of the above

8. A test case contains
   a. input with the expected output.
   b. output with the expected input.
   c. input with actual output.
   d. None of the above

9. A test suite is
   a. the collection of test cases used to check the software.
   b. the collection of errors.
   c. protocol.
   d. infinite.

10. Which of the following technique is not related to software testing?
    a. Random testing
    b. Symbolic testing
    c. Physical testing
    d. Metaheuristic testing

SHORT-ANSWER QUESTIONS

1. Write down the finding of the no-free-lunch theorem.
2. What is random testing?
3. What is a flow graph?
4. Is complete testing of software possible?
5. What is black-box testing?
6. What is white-box testing?
7. Explain the working of the GCOV tool.
8. Write down the difference between the GCOV and TRUCOV tools.
9. Explain symbolic testing.
10. Explain dynamic testing.

LONG-ANSWER QUESTIONS

1. Explain how a GA can be used to generate test cases for a program. Take an example program and explain.
2. Explain how PSO can be used to generate test cases for a program. Take an example program and explain.
3. Other than metaheuristic techniques, what other methods can be used for test case generation?
4. Is only one test case sufficient to cover the entire software? If not, explain the method of testing software using DE.
5. Why a single optimization algorithm cannot be used in solving all the problems? Why will it not be effective?
6. Discuss how a test case can improve the quality of the software. Will all test cases improve the quality of software?

CHAPTER 10

MULTIPLE-CHOICE QUESTIONS

1. What is required to check by the companies before starting software testing?
    a. Time and cost available for testing
    b. Only time
    c. Only cost
    d. None of the above

2. Regression testing is done at
   a. the time of software testing.
   b. the time of maintenance.
   c. Both a and b
   d. None of the above

3. Which of the following is not a part of regression testing?
   a. Test case prioritization
   b. Test case minimization
   c. Test case selection
   d. None of the above

4. Which version of GA can be used for test suite minimization?
   a. The real-parameter version
   b. The binary version
   c. Both a and b
   d. None of the above

5. If the size of a test suite is 100, how many subsets of test cases can be created?
   a. 100
   b. 200
   c. 2^{100}
   d. $2^{100} - 1$

6. Why is test case prioritization required?
   a. For saving time and cost
   b. Only time
   c. Only cost
   d. None of the above

7. What will be the size of the initial population if we want to minimize a test suite with 100 test cases?
   a. 100
   b. 40
   c. 100 * 40
   d. Decided by the programmer

8. In Question 7, if a binary GA is applied, how many bits will be there in a chromosome?
   a. 40
   b. 100
   c. 200
   d. 50

LONG-ANSWER QUESTIONS

1. Explain the significance of regression testing in software industries?
2. How we can reduce the size of the test suite, keeping its efficiency high in detecting the faults in the software?
3. Apart from GA, PSO, and DE, what are the other nature-inspired algorithms that could be applicable in the test suite minimization?

CHAPTER 11

MULTIPLE-CHOICE QUESTIONS

1. Which of the following cloud supplier provides IaaS?
   a. Amazon EC2
   b. Microsoft Azure
   c. Both a and b
   d. None of the above

2. A workflow defines
   a. task-to-task mapping.
   b. task-to-VM mapping.
   c. VM-to-VM mapping.
   d. None of the above

3. Transfer cost is associated with
   a. the cost required to execute the task on a VM.
   b. the cost required to execute workflow on a VM.
   c. the cost required to transfer the task from one VM to another.
   d. the cost required to transfer workflow from one VM to another.

4. A fitness function used to optimize the cost of workflow includes
   a. the transfer cost.
   b. the transfer cost and the execution cost.
   c. the execution cost.
   d. None of the above

5. A penalty may be added in the fitness function
   a. when heterogeneous VMs are used.
   b. when homogeneous VMs are used.
   c. when load balancing is required.
   d. None of the above

SHORT-ANSWER QUESTIONS

1. Define *cloud computing*.
2. What is the difference between an IaaS cloud and a PaaS cloud?

3. Define *workflow*.
4. What is a workflow scheduling problem? Explain with one example.
5. Show how workflow is represented in a PSO algorithm.

LONG-ANSWER QUESTIONS

1. Is it possible to apply a differential evolution algorithm to a workflow scheduling problem? If yes, explain how the solution will be represented.
2. Define binary encoding for representing a workflow schedule of five tasks on eight VMs.
3. Is it possible to minimize the time for executing a workflow schedule? Who will benefit from such an application?
4. How will a fitness function be designed to minimize the time of a workflow schedule?
5. Other than workflow scheduling problems, identify other optimization problems in the cloud scenario.

Answers

CHAPTER 1

1. a: traveling salesperson problem.
2. b: a finite number of solutions.
3. c: the fitness value of a solution is evaluated.
4. b: it is adjustable with dimension.
5. b: possibly different.
6. c: both a and b
7. c: Both a and b.
8. b: Multimodal problem.
9. a: Unimodal problem.
10. c: Both.

CHAPTER 2

1. b: the theory of natural selection and evolution.
2. b: to promote good solutions.
3. c: Sharing Fitness method.
4. b: Schemata theory.
5. a: To create new solutions.
6. c: Both a and b.
7. c: Both a and b.
8. d: The number of solutions used to start the search.
9. b: Chromosome.
10. d: All.
11. c: Rank-based selection.
12. d: both a and b.
13. b: the fitness of the solutions.
14. c: One point.
15. a: True.
16. a: True.
17. c: Rank.
18. a: the number of variables in the solution.
19. d: Both b and c.
20. d: 40.

CHAPTER 3

1. b: No.
2. a: Yes.
3. c: of size 3×3 having random numbers between 0 to 2.

4. d: All of the above.
5. b: No.

CHAPTER 4

1. b: DE algorithm.
2. c: mutation, crossover, and selection.
3. d: *F*.
4. b: from the target vector and the trail vector.
5. b: donor vector.
6. d: All of the above.
7. a: numerical optimization problems.
8. b: Target vector.
9. c: One-point crossover.
10. a: every parent solution.

CHAPTER 5

1. b: Bird flocking.
2. b: For implementing exploitation in a PSO.
3. a: For implementing exploration in PSO.
4. c: Clamping.
5. c: the fitness of the new position is better than the previous position.

CHAPTER 6

1. c: hunting process used by grey wolves.
2. b: swarm intelligence algorithm.
3. a: checking the fitness of each wolf in the population.
4. c: To work according to the instructions received from alpha, beta, and delta wolves.
5. b: No.

CHAPTER 8

1. c: SA.
2. c: Cuckoo search algorithm.
3. c: CMA-ES algorithms.
4. a: Ant–bee colony.
5. d: Bat algorithm.

CHAPTER 9

1. c: Both a and b.
2. c: Symbolic testing.

3. d: Any of a, b, and c.
4. c: Both a and b.
5. a: A test case that produces the expected output.
6. b: Black-box testing.
7. a: Test cases will be generated to meet the requirement of the software.
8. a: input with the expected output.
9. a: the collection of test cases used to check the software.
10. c: Physical testing.

CHAPTER 10

1. a: Time and cost available for testing.
2. a: the time of software testing.
3. d: None of the above.
4. b: The binary version.
5. c: 2^{100}.
6. a: For saving time and cost.
7. d: Decided by the programmer.
8. b: 100.

CHAPTER 11

1. a: Amazon EC2.
2. b: task-to-VM mapping.
3. c: the cost required to transfer the task from one VM to another.
4. b: the transfer cost and the execution cost.
5. c: when load balancing is required.

Index

A

D

F

G

J

K

L

N

U

V

Y

Z